ESCAPISM

Published in
Cooperation with
the Center for
American Places,
Harrisonburg,
Virginia

Other Books by Yi-Fu Tuan

Pediments in Southeastern Arizona

The Hydrologic Cycle and the Wisdom of God

China

Man and Nature

Topophilia: A Study of Environmental Perception, Attitudes, and Values

Space and Place: The Perspective of Experience

Landscapes of Fear

Segmented Worlds and Self: Group Life and Individual Consciousness

Dominance and Affection: The Making of Pets

The Good Life

Morality and Imagination: Paradoxes of Progress

Passing Strange and Wonderful: Aesthetics, Nature, and Culture

Cosmos and Hearth: A Cosmopolite's Viewpoint

Yi-Fu Tuan

ESCAPISM

The Johns Hopkins University Press

• • •

Baltimore & London

Printed in the United States of America
on acid-free paper

9 8 7 6 5 4 3 2

The Johns Hopkins University Press
2715 North Charles Street
Baltimore, Maryland 21218-4363
www.press.jhu.edu

The Johns Hopkins Press Ltd., London

Library of Congress Cataloging-in-Publication
Data will be found at the end of this book.
A catalog record for this book is available
from the British Library.

ISBN 0-8018-5926-3

To Betty Parsons

and in the memory of

Jim Parsons

Contents

• • •

Preface

• • •

Who hasn't—sometime—wanted to escape? But from what? To where? And once we have arrived at the good place, is this the end of the desire to move? Or does it stir again, tempted by another image, even if it be that of the place from which we started—our old home or childhood? Surely everyone has had the urge to be elsewhere in moments of stress and uncertainty. I have. Yet even though I am a geographer, whose business is to study why and how people move, as well as their ceaseless striving to make wherever they are an even better place, the word "escape" or "escapism" rarely came to the forefront of my consciousness, never offered itself as a possible key to the understanding of human nature and culture.

All this changed unexpectedly a few years ago when I received an invitation to write a paper on the landscape of Disneyland. Not being an expert on theme parks, I was initially inclined to say no. But I live in Madison, Wisconsin. The call came in November, and the caller said that prospective authors would be asked to meet in Anaheim, in southern California, where Disneyland is located, in January. Suddenly the word "escape" flashed across my mental screen: I could shake off the grip of Wisconsin winter for a restorative spell in Lotus Land. I accepted. And how glad I am now to have done so, for not only did I enjoy my visit, it also prompted me to raise questions, the tentative answering of which has produced this book.

I enjoyed my visit, if only because the sunny warmth was most welcome. But, to my surprise, I found Disneyland itself delightful. I say "to my surprise" because well-educated people, among whom I count myself, are taught to dismiss the theme park as an unreal, fantasy world supported by hidden—and therefore somewhat sinister—forces. My unexpected response led me to ask a series of questions. Granted that theme parks are escapist fantasies, suitable only for the immature, what human works aren't? Is there a ladder of aspiration or pretension, at one end of which are the exuberantly or crassly playful and at the other end the deeply serious and real?

Suppose I move down the ladder. What comes after theme park? Shopping mall? It has been attacked as an escapist Eden for mindless consumers. Suburb? Academic detractors have not hesitated to dismiss it as a dull, middle-class playground. They prefer the city. But the city is escapist par excellence, for a city is a city—a real city!—to the degree that it has distanced itself (escaped) from nature and its rhythms. Is farm life, being so close to nature, the ultimately real? Urban sophisticates in a nostalgic mood seem to think so. Yet farmers have obviously striven to create their own world, and in any proud farmhouse, pictures hang on the wall, artificial light drives out darkness. Hunter-gatherers? They have barely modified their natural environment. They don't have the tools. But they do have the tool of language, and with it they, like all humans, have woven an alternative or complementary reality to which they can resort for support in times of stress and in which they can take delight.

Seeing culture as escapism raises the question, Escape from what? My background as a geographer predisposes me to consider, first, escape from nature—from the natural environment, its uncertainties and threats. Success in modern times has introduced unprecedented predictability and plenitude into human life. This ought to ensure happiness, but it doesn't; it doesn't even ensure deep security. Modern men and women, living in their artificial worlds high up on the ladder of aspiration and pretension, seem to suffer from what Milan Kundera calls the "unbearable lightness of being." Life up there doesn't seem quite real—there again, that problematic word—and although people do not mind living in a pleasant dream, they might well think something amiss living and dying in one without ever knowing what it is to be awake. Hence the desire to pause on the ladder of aspiration, or even to come down a rung or two. Extremists yearn to sink to the bottom, to hug the earth, the gritty texture and harshness that make for reality. Moderates seek a middle position on the ladder, called by geographers and other students of the environment the middle landscape.

Nature, to geographers, is the external natural environment. Where does that leave the body—in particular, *my* animal body? The body is clearly nature. But it is not external to me. It is me: I don't want to, and in any case I can't, escape from it. Yet this isn't quite true. In pain, I have often wished that I could abandon my body and be elsewhere. It is even possible to do so to a limited degree. I can always resort to imagination, which is the most readily available means of transporting the self, momentarily, out of its body. If external nature is often unsatisfactory from the viewpoint of my needs and aspirations, so is the nature that is my body. I meddle with it, and much of my meddling comes out of a conscious desire to escape my animality. An animal eats, has sexual drives, and sooner or later dies. I? Well, I dine, love, and aspire to be immortal. Culture is the totality of means by which I escape from my animal state of being.

This "I" is a mixture of East (Chinese) and West. My dual background has made me curious to see how they compare—in particular, how Chinese and Westerners have sought, historically, to distance themselves from nature, including their animal nature. In my view, the West has pushed further toward artificiality than has the East. Does this mean that elite Westerners are more civilized, more sophisticated, than elite Chinese? Or are they just higher up the ladder of pretension, more out of touch with the real?

"I" may be a term of convenience for any human individual—the proudly universal "I" and Everyman. "I" may also stand for the individual representative of a group: "I" as in I am a Chinese-American. This can be a source of pride too, pride in one's group and its values. Then there is the very specific "I"—for example, the author of the present book. I am different from any other individual. It is good to be different. I am proud to be unique. Yet at a deeper level, being different, unique, is intolerable. It makes for disconnectedness, meaninglessness, loneliness, vulnerability. Immersing oneself in a larger entity—the group—is a compelling human need. When I discussed this book with a friend, she was sympathetic to the general thesis. She said, "People do indeed want to escape, opt out, sometimes through desperate means. But don't they want stability even more? Don't they want to be in place, to be part of an unchanging whole?" Yes, I say, of course they do. A major task of culture is to promote order and stability. Rules and regulations have no other purpose. The material environment itself signifies stability and, moreover, makes it easier for an individual to feel, as she looks at this or that familiar thing in a familiar place, that she belongs. There are different ways to escape, but if what one wishes to escape from is the singularity, frailty, and openness to change of the self, then there is no better way than immersion in the group and its numerous anchoring devices.

Culture is driven by imagination and is a product of imagination. We humans are pleased and proud to have it. But imagination can lead us astray—into fantasy, the unreal, and the grotesque; and it can tempt us into first picturing, then (too often) acting out evil. Even the good that imagination brings forth can be ambivalent. It is good to escape from raw and threatening nature into the refinements of culture. However, viewed through darker glasses, these refinements have the appearance of a towering froth burying and hiding the harsh economic and political realities that sustain it and made it possible in the first place. Covering up, distancing, escaping, makes it easy for us to forget the destructive preliminaries of almost all creative acts—even one as basic and necessary as cooking. Few cookbooks bear the title *Butchering and Cooking* or *Evisceration and Cooking*, and yet how is the one possible without the other? And if the project is not food but a monument, a city, an empire, the amount of prior destruction, the exploitation of labor both animal and human, the sweat, pain, and death, add up to a picture closer to hell than heaven. Moreover, even if we remove the dark glasses and just concentrate on the sparkling artifactual world we have built, living in one, as I have noted, can feel curiously lightweight and unreal.

Nevertheless, to put it thus is to tell a very incomplete and biased story. If the emergence of life and sensation in the universe may be considered a good, the emergence of human life and imagination may be considered another good—a greater good. Historically, imagination in its highest flights has led some bold spirits not into solipsistic fantasy and madness, or into evil, but into genuine encounters with the sublimities of the physical universe, encounters that can have a moral effect—for example, a greater sense of humility. At a more mundane level, imagination is constantly at work, enchanting and reenchanting the world; that is to say, it enables people to see import (magic and beauty) in nature and humanly made things hitherto unrecognized, or recognized and then forgotten. One rather striking example is "landscape." Regarded fondly by wealthy Romans in antiquity and by practically everyone nowadays, it was largely unknown to Europeans in the period from 400 to 1400. The Renaissance saw the rebirth of a sensibility, the reenchantment of land as landscape, that has persisted, with alterations and improvements, to our time. I am tempted to say that "landscape" is congenial to our spirit, that once it is pointed out to us, we not only nod in recognition but feel better, healthier, empowered.

A world, however enchanted, is in the end frivolous without moral weight. What does it mean to be good? What does "good" mean? An entire history of human striving lies behind these questions, the tentative answering of which—and, far more important, the tentative realization of the "good" so discovered and articulated—is the closest we, in this life, can get to heaven.

Escapism, I will argue, is human—and inescapable. There is nothing wrong with escape as such. What makes it suspect is the goal, which can be quite unreal. And what is wrong with the unreal—with wild fantasy? Nothing, I would say, so long as it remains a passing mood, a temporary escape, a brief mental experiment with possibility. However, fantasy that is shut off too long from external reality risks degenerating into a self-deluding hell—a hell that can nevertheless have an insidious appeal. The transposition of this personal hell to others, through naked power or cunning rhetoric, is a great evil. But escape can just as well be in the opposite direction—toward the real and the good (heaven). To skeptical modern men and women, such a happy move—on the face of it, as plausible as the move in the other direction—rings false. "Heaven," for them, is almost another word for delusion and fantasy. If anything is real, it is the harshness and brutality of life, or hell. I wish to counter this fashionable pessimism.

In writing this book, I have two broad aims. One is to provide an unusual and, I believe, fruitful perspective on nature and culture. My other aim is to persuade readers, especially those who have fed too exclusively on the literature of despair, to recognize the good that has already been—though insecurely—achieved, and hence to look upon the idea, if not of heaven on earth, then of an earth periodically visited by heavenly bliss, in a less dismissive, more hopeful, light.

ESCAPISM

1

Nature and Culture

• • •

"Escapism" has a somewhat negative meaning in our society and perhaps in all societies. It suggests an inability to face facts—the real world. We speak of escapist literature, for instance, and we tend to judge as escapist places such as mega–shopping malls, fancy resorts, theme parks, or even picture-perfect suburbs. They all lack—in a single word—weight.[1]

Suspicion of escapism has many causes. The most obvious is that no animal can survive unless it perceives its environment as it really is. Daydreaming or wishful thinking would not answer. The hard facts cannot be made to go away by shutting one's eyes. But so far as we know, only humans may withdraw, eyes shut, to ponder the nature of a threat rather than confront it directly, muscles tense, eyes open; only they daydream and engage in wishful thinking. Significantly, only humans have culture. By culture I mean not just certain acquired habits, the manufacture and use of certain tools, but a whole world of thought and belief, habits and customs, skills and artifacts. Culture is more closely linked to the human tendency not to face facts, our ability to escape by one means or another, than we are accustomed to believe. Indeed, I should like to add another definition of what it is to be human to the many that already exist: A human being is an animal who is congenitally indisposed to accept reality as it is. Humans not only submit and adapt, as all animals do; they transform in accordance with a preconceived plan. That is, before transforming, they do something extraordinary, namely, "see" what is not there. Seeing what is not there lies at the foundation of all human culture.

Reality and the Real

What do the words "reality" and "real" mean? Although philosophers do not find it easy to agree on an answer, ordinary thinking people have little difficulty using these words in everyday talk, often in conjunction with their opposites, "fantasy" and "unreal." Such talk, when looked at closely, shows how the meaning of "real" shifts, even radically, as the context changes. A common meaning draws on the model of animal life. The idea is that animals live in the real world, respond as best they can to outside forces and their own nature, free of unsettling images and aspirations. Humans can approach that state of existence by also living close to nature, curbing the imagination and jettisoning excess cultural baggage. Nature itself is real. It is indubitably real to humans when they feel it as a blast of cold wind, a sudden shower, or the skin rash caused by contact with poison ivy. So another meaning of "real" emerges: the real as impact. It is not just nature; it is whatever in nature or in society imposes itself on a human being or group, doing so either suddenly or as a consistently felt pressure. "Reality" in this sense is intractable, and it is indifferent to the needs and desires of particular individuals and groups. Facing reality, then, implies accepting one's essential powerlessness, yielding or adjusting to circumambient forces, taking solace in some local pattern or order that one has created and to which one has become habituated. This "local pattern or order" points to another sense of the real: a small and thoroughly humanized world. Far from being shock or impact, the real is the familiar, the predictable, the nurturing and all-enveloping. Home is the prime example. Home is a place to which one is attached by myriad habits of thought and behavior—culturally acquired, of course, yet in time they become so intimately woven into everyday existence that they seem primordial and the essence of one's being. Moving out of home and the familiar, even when this is voluntary and of short duration, can feel like escapism, sojourn in a fantasy world, less real because less dense and all-encompassing.

Does this conclude the list of commonly accepted meanings of "real"? No. For completion, at least one more sense of the word demands to be added. Disconcertingly, it is the opposite of the one I have just given. In this usage, it is daily life, with its messy details and frustrating lack of definition and completion—its many inconclusive moves and projects twisting and turning as in a fitful dream—that is unreal. Real, by contrast, is the well-told story, the clear image, the well-defined architectural space, the sacred ritual, all of which give a heightened sense of self—a feeling of aliveness.

The Earth

The earth is our home. Trips to the moon, another planet, a distant star, have haunted the human imagination and may even become a commonplace reality one day. But they nevertheless have an aura of fantasy about them. Real life is life on earth; it is here that we have our roots and our being. Geographers study the earth as human habitat or home. Interestingly, they discover that the earth is never quite the home humans want it to be; hence the dreams of flying and of a paradise located elsewhere that are common to many cultures. Most people, when they think of the earth, think not of the entire planet but of a part of it—the part they live in. Wherever they happen to be, provided they have been settled there for some time, they consider home. Yet this is not quite the case either, if only because if it were, there would be no story, no *human* story, to tell; people, like other animals, will be "immersed in nature," as G. W. F. Hegel put it. It is the restless activity that produces the story line. Human beings have been and continue to be profoundly restless. For one reason or another, they are not content with being where they are. They move, or if they stay in place, they seek to rearrange that place. Migration and the in situ transformation of the environment are two major themes—*the* two major themes—in human geography. They both reveal a discontent with the status quo, a desire to escape. Geographers have written voluminously on these themes without using "escape" or "escapism" as a guiding concept. What is to be gained by using it now? The gain is that it forces us to reconsider nature and culture, and thereby who we are and what we aspire to, in productive tandem with "real and imagined," "reality and fantasy"—ideas that traditionally are at the core of humanist scholarship and thinking.

Migration

Migration is clearly a type of escape. Animals move out when their home ground starts to deteriorate. Humans have done so since the earliest times; and it now appears that as they acquired certain critical marks of culture—outstandingly, language around sixty thousand years ago—they became better able to organize themselves in complex ways and meet the challenges of the environment by migrating, sometimes over great distances. To overcome great distance, our remote ancestors must have had not only organizational ability, enormously enhanced by language, but also new technical means at their disposal—seaworthy craft, for example. Such people must have been of lively mind and were, I will assume, quite capable of envisaging "greener pastures" elsewhere and making plans as to how best to reach their destination.[2] By the end of the Ice Age, some twelve thousand years ago, human beings had spread into every kind of natural environment, from the Tropics to the Arctic, the major exceptions being ice sheets and the highest mountains.

Much of the human story can be told as one of migration. People move a short distance to a better hunting ground, richer soil, better economic opportunity, greater cultural stimulus. Short-distance movements are likely to be periodic, their paths winding back on themselves with changing circumstance. Over the years such movements become habit, their circuits habitat. Long-distance migrations, by contrast, are likely to be in one direction and permanent. A certain epic grandeur attaches to them, for migrants must be willing to take steps that make life even more difficult than it already is in the hope of future felicity. Before people make a risky move, they must have information about their destination point. What kinds of information are available? To what extent does the need to believe in a better world at the horizon overrule or distort the "hard facts" that people know? Is reality so constraining and unbearable at home that it becomes the seedbed for wild longings and images? And do these images, by virtue of their simplicity and vividness, seem not a dream but more "real" than the familiar world? A great modern epic of migration is the spread of Europeans to the New World. The United States of America proclaims itself a land of immigrants. It would not want to be known as a "land of escapists," yet many did just that: escape from the intolerable conditions of the Old World for the promises of the New.[3]

Nature and Society

Human restlessness finds release in geographical mobility. It also finds release (and relief) in bringing about local change. The circumstance one wishes to change—to escape from—can be social, political, or economic; it can be a run-down urban neighborhood or a ravaged countryside. And it can be nature. In telling a human story, we may start at any point in time, but if we go back far enough we necessarily have nature, untouched nature, as stage: first the swamp, forest, bush, or desert, then . . . then what? Then humans enter, and our story begins.

In the long run, humans everywhere experience, if not forthrightly recognize, nature as home and tomb, Eden and jungle, mother and ogre, a responsive "thou" and an indifferent "it." Our attitude to nature was and is understandably ambivalent. Culture reflects this ambivalence; it compensates for nature's defects yet fears the consequences of overcompensation. A major defect is nature's undependability and violence. The familiar story of people altering nature can thus be understood as their effort to distance themselves from it by establishing a mediating, more constant world of their own making. The story has many versions. Almost all are anguish-ridden, especially early on, when pioneers had to battle nature for a precarious toehold.[4]

A natural environment can itself seem both nourishing and stable to its human habitants. A tropical forest, for example, provides for the modest needs of hunter-gatherers throughout the year, year after year. However, once a people start to change the forest, even if it is only the making of a modest clearing for crops and a village, the forest can seem to turn into a malevolent force that relentlessly threatens to move in and take over the cleared space.[5] Some such experience of harassment is known to villagers all over the world, though perhaps not to the same degree as in the humid Tropics. Villagers are therefore inclined to see nature in a suspicious light. Of course they know that it provides for their needs and are grateful—a gratitude expressed by gestures and stories of respect. But they also know from hard experience that nature provides grudgingly, and that from time to time it acts with the utmost indifference to human works and lives.

Carving a space out of nature, then, does not ensure stability and ease. To the contrary, it can make people feel more than ever vulnerable. What to do? Lacking physical power, the most basic step they can take is to rope nature into the human world so that it will be responsive—as difficult people are—to social pressures and sanctions. If these don't work, they try placatory ritual, and if this in turn fails, they appeal to the higher authority of heaven or its human regents on earth. By one means or another they seek control, with at best only tenuous success. What appears stable to the visiting ecologist, whose discipline predisposes him to focus on long-range people-environment interactions, may be not stable at all but rather full of uncertainty to the local inhabitants struggling to survive from day to day, week to week, one season to another.

Aztec and Chinese

Now, suppose we turn to a more advanced society, one in which the people, unlike isolated and poorly equipped villagers, have the technical and organizational means to make extensive permanent clearings, raise crops, and build monuments, including cities. Won't the exercising of that power and the looming presence of large human works impress on them a sense of their own efficacy and the world's permanence?

The answer is, Not always. The Aztecs of Mexico are a case in point. Here is a people who continued to feel insecure despite the scope and sophistication of their material attainments. Nature's instabilities, made evident by the ominous presence of volcanoes and experienced repeatedly in the wayward behavior of weather, stream flow, and lake level, more than overruled whatever reassurance human artifacts could give. Moreover, in the Aztec civilization the architectural monuments of temple and altar themselves attested more to fear and anxiety than to confidence, for they were built to conduct human sacrifice, with the end in view of sustaining and regulating the enfeebled forces of the cosmos.[6]

Consider another, more confident civilization, the Chinese. The Chinese struggled to regulate nature through physical intervention and by such institutional means as the establishment of public granaries. In these respects the two civilizations, Aztec and Chinese, had something in common. However, unlike the Aztecs, the Chinese managed to sustain over the course of millennia, and in the teeth of abundant contrary evidence, a magnificent model of cosmic harmony. This ability to overlook evidence may earn the Chinese the label of escapists, but without it—without their tenacious hold on the dream of harmony—they would have deprived themselves of optimism and fortitude, psychological advantages that helped them to create an enduring culture.[7] To the Chinese architect-engineer, barriers such as swamp, forest, and hillock could be overcome; they did not have to be accepted as embossed in the eternal order of things. And to the Chinese philosopher—indeed, to the philosopher in any culture—wayward facts and contingencies were not just there to be noted and accepted; rather, they were puzzling pieces of reality that could stimulate one to search for a more comprehensive world-view.

Chinese composure has its source in a number of factors. Tangible architectural and engineering achievements no doubt promoted confidence, as did the memory of extended periods of peace and prosperity during a great dynasty such as the Han, T'ang, or Sung. The Chinese inclination to see the universe as orderly and hence accessible to reason surely also promoted composure. Even more reassuring—more wishful and escapist, from our secularist-modern perspective—is the idea that the universe is moral and hence responsive to moral suasion. China has had its share of natural disasters; these might well have been more devastating and frequent than those that afflicted the Aztec empire. When disasters visited China and could not be alleviated by ordinary means, the emperor took responsibility, for he considered them to be a consequence of his own moral failing. To reestablish order, he "memorialized heaven," imposing a penance on himself on his own and humankind's behalf. As exemplary man, the emperor was the ultimate mediator between heaven and earth; and for this reason he could by his own conduct and sacrifice right wrong, restore harmony throughout the worlds of nature and of people.[8] The emperor was called Son of Heaven rather than Son of Earth. There is no doubt a heavenward tilt in Chinese high culture, as there is in all high cultures. What high culture offers is escape from bondage to earth.

Premodern and Early Modern Europe

Escape from nature's vagaries and violence—except during a blizzard or hurricane—may seem a strange idea to modern Westerners, for whom society rather than nature is unpredictable and violent. How short is their memory! Any full account of life and livelihood in the West from the Middle Ages to the eighteenth century must give a prominent role to weather—that is, if we are more interested in ordinary people than in potentates and their political high jinks. So much misery had its immediate cause in meteorological freakishness. Too much rain or too little, prolonged cold or withering heat, led to crop failure and, all too often—at least locally—to famine and starvation.

Records from the early modern period show how frequently people even in the richest parts of Europe suffered and died from lack of food. In 1597 a citizen of Newcastle wrote of "sundry starving and dying in our streets and in the fields for lack of bread." And this despite importation of foreign grain into the port city. In France, rather than the sort of cosmic stability to be expected from a Sun King, wild swings of lean and fat years seemed more the rule. In 1661–62 much of France was afflicted by bad weather, poor harvests, and famine. Beggars from the countryside flocked to the towns, where citizens formed militias to drive them back. Good weather produced good harvests in 1663; there followed a decade of prosperity. From 1674 onward, however, the times were once more "out of joint." A wet summer curtailed the harvests of 1674; those of 1677, 1678, and 1679 were worse. Yields were again poor in 1681 and catastrophic in some regions in 1684. Between 1679 and 1684 the death toll rose throughout much of France. Good weather prevailed from 1684 to 1689; magnificent harvests made for cheap grain, and the people were, for a change, more than adequately fed. Then came the great famine of 1693–94, the culmination of a succession of cold and wet years. A majority of people in France suffered, though in varying degree. Poor folks resorted to eating "such unclean things as cats and the flesh of horses flayed and cast on to dung heaps," and some starved to death.[9]

In premodern and early modern Europe, uncertainty in both nature and society put a heavy burden on the poor. That hardly surprises us. More difficult for us to imagine now is how uncertainty could haunt the well-to-do, even the rich and the powerful.[10] When uncertainty is so much a fact of life, escape into a make-believe world of perfect order may be excused. Make-believe was one way—an important way—that Renaissance princes coped. They produced elaborate masques in which they themselves sometimes played the roles of gods and goddesses reigning in a pastoral heaven of abundance and peace. If ordinary people sought to exclude unruly weather by putting a roof over their heads, Renaissance rulers did that and far more. By means of the art at their command they produced an alternative heaven: the palace itself and, even more overtly, the theatrical stage of floating clouds, flying chariots, pastures and billowing fields of surpassing fertility.[11]

What was the nature of this art? Shakespeare hinted at it in the magic powers of Prospero. A Renaissance prince was a Prospero—a magician. A magician was not the marginal entertainer we now see him to be. Rather, he was considered a person of deep knowledge—someone who knew how things worked below the surface and so could do wonders. Whereas a prince only purchased such power, a genius like Leonardo da Vinci possessed it in his own person to a remarkable degree. It doesn't seem to me far-fetched to call Leonardo a magician. Indeed, a much later figure, Isaac Newton, has been called a magician, the last one. An important difference, however, separates a Renaissance figure like Leonardo and the outstanding genius of a later time, Newton: Leonardo approached knowledge through art, technique, and technology, skills that would have been necessary to the making of the sort of surrogate heaven that Renaissance princes yearned for. By contrast, Newton showed little interest in the earthbound phenomena such as anatomy and geology that fascinated Leonardo. Nor was he concerned with building a surrogate heaven on earth in the manner of an artist-architect. Rather, his gaze was directed to heaven itself, and his singular contribution to knowledge was through the abstractions of mathematics.

Physical vs. Biological Science: Heaven vs. Earth

Alfred North Whitehead, an outstanding mathematician-philosopher of our time, famously designated the seventeenth century the Century of Genius. He gave twelve names: Bacon, Harvey, Kepler, Galileo, Descartes, Pascal, Huyghens, Boyle, Newton, Locke, Spinoza, and Leibniz. He apologized for the predominance of Englishmen, then noted without apology that he had only one biologist on the list: Harvey.[12] Genius in that century showed itself in celestial mechanics and physics rather than in biology or organic nature, to which humans belong and upon which they depend. At the threshold of the modern age, human helplessness—the recurrent famines and starvation I mentioned earlier—continued to exist even in the developed parts of Europe; on the other hand, the laws of celestial nature were being mapped with unprecedented accuracy. On earth, both nature and human affairs often seemed to verge on chaos; heaven, by contrast, exhibited perfect order. Cosmic order gave the natural philosophers of the seventeenth century confidence, as it has given confidence to priest-kings throughout human history. In premodern times, rulers believed that the regularities discernible above could somehow be brought down below. In early modern Europe, natural philosophers had grounds for hoping that the rigorous method that opened the secrets of heaven could work similar wonders on earth. For two centuries, however, there was hardly any link between the splendid theoretical reaches of the new science and applications that catered to ordinary human needs. Agricultural advances during the eighteenth century had more to do with changes in practice (crop rotation, for example), in a more systematic use of knowledge gained through centuries of trial and error, in changes of land tenure and ownership, and suchlike than with the bright discoveries of an abstract, mechanistic, heaven-inspired science.

To this observation a critic may say, "Well, what do you expect? The challenges of agriculture can be met only by close attention to the intricacies and interdependencies of land and life, to what is happening at our feet and before our eyes rather than in a scientist's playpen (the laboratory), or by seeking models of analysis and conceptualization suited to the world of astronomy and physics. In short, to live well, one needs more down-to-earth realism, not escapism."

This sensible answer has its own difficulties. As we now know, what may be deemed escapism turns out to be a circuitous route to unprecedented manipulative power over organic life, and not just predictive power over the stars. One branch of the route took the scientific and entrepreneurial spirits of the West from the study of general chemistry to the study of soil chemistry, and from there to the manufacture of chemical fertilizers, the use of which led to impressively higher crop yields; another branch took them from the study of genetics to the scientific breeding of plants and animals, which became more and more ingenious, reaching a high peak in the Green Revolution, and onward to genetic engineering. While all this was taking place, the same theoretic-analytic bent of mind produced agricultural machines of increasing power and flexibility, and, one might add, the complex organizational and marketing strategies of advanced farming. Countries that embraced these discoveries and inventions prospered. In the second half of the twentieth century, cornucopia no longer seems just a dream, as it has been for the vast majority of people throughout human time. A substantial number of people in the developed parts of the world encounter it day after day. They have learned to take the supermarket's dazzling pyramids of fruits and vegetables, its esplanades of meat, for granted. And yet a doubt lingers as to whether such abundance is real and can last, whether it is not just an effect of Prospero's magic wand. The upward curve of success in the West has not altogether removed the feeling that technological society must, sooner or later, pay for its arrogation of powers that rightly belong only to nature and nature's God.

Escape to Nature

I have given a brief and sweeping account of "escape from nature," which has taken us from uncertain yields in village clearings to supermarket cornucopia. The escape is made possible by different kinds of power: the power of humans working cooperatively and deliberatively together, the power of technology, and, underlying them, the power of images and ideas. The realities thus created do not, however, necessarily produce contentment. They may, on the contrary, generate frustration and restlessness. Again people seek to escape—this time "back to nature."

Escaping or returning to nature is a well-worn theme. I mention it to provide a counterpoint to the story of escaping *from* nature, but also to draw attention to certain facets of the "back to nature" sentiment that have not yet entered the common lore.

One is the antiquity of this sentiment. A yearning for the natural and the wild goes back almost to the beginning of city building in ancient Sumer. A hint of it can already be found in the epic of Gilgamesh, which tells of the natural man Enkidu, who was seduced by gradual steps to embrace the refinements of civilization, only to regret on his deathbed what he had left behind: a free life cavorting with gazelles.[13]

The second point I wish to underline is this: Although a warm sentiment for nature is common among urban sophisticates, as we know from well-documented European and East Asian history, it is not confined to them. The extreme artificiality of a built environment is not itself an essential cause or inducement. Consider the Lele of Kasai in tropical Africa. They do not have cities, yet they know what it is like to yearn for nature. What they wish to escape from is the modestly humanized landscape they have made from the savanna next to the Kasai River, for to keep everything there in good order—from social relations to huts and groundnut plots—they must be constantly vigilant, and that proves burdensome. To find relief, the Lele men periodically leave behind the glare and heat of the savanna, with its interminable chores and obligations, to plunge into the dark, cool, and nurturing rain forest on the other side of the river, which to them is the source of all good things, a gift of God.[14]

The third point is that "back to nature" varies enormously in scale. At one end of the scale are such familiar and minor undertakings as the weekend camping trip to the forest and, more permanently, the return to a rural commune way of life. At the other end of the scale is the European settlement of North America itself. It too might be considered a type of "escape to nature." Old Europe was the city; the New World was nature. True, many settlers came from Europe's rural towns and villages rather than from its large cities. Nevertheless, they were escaping from a reality that seemed too firmly set and densely packed to the spaces and simpler ways of life in the New World.[15]

My final point is this: Back-to-nature movements at all scales, including the epic scale of transatlantic migration, have seldom resulted in the abandonment, or even serious depletion, of populations in the home bases—the major cities and metropolitan fields, which over time have continued to gain inhabitants and to further distance themselves from nature.

This last point serves to remind us that "escape to nature" is dependent on "escape *from* nature." The latter is primary and inexorable. It is so because pressures of population and social constraint must build up first before the desire to escape from them can arise; and I have already urged that these pressures are themselves a consequence of culture—of our desire and ability to escape from nature. "Escape from nature" is primary for another reason, namely, that the nature one escapes to, because it is the target of desire rather than a vague "out there" to which one is unhappily thrust, must have been culturally delineated and endowed with value. What we wish to escape to is not "nature" but an alluring conception of it, and this conception is necessarily a product of a people's experience and history—their culture. Paradoxical as it may sound, "escape to nature" is a cultural undertaking, a covered-up attempt to "escape from nature."

Nature and Culture

Nature is culturally defined, a point of view that is by now widely accepted among environmental theorists.[16] Culturally defined? Humanly constituted? Is this the latest eruption of hubris in the Western world? Not necessarily, for the idea can reasonably be coupled with another one, inspired by Wittgenstein, namely this: That which is defined and definable, that which can be encompassed by language or image, may be just a small part of all there is—Nature with a capital *N*.[17] Now, in this chapter I myself have been using the word "nature" in a restricted sense—nature with a small *n*. What do I mean by it? What is the culture that has influenced me? It is the culture of academic geography. The meaning that I give the word is traditional among geographers: Nature is that layer of the earth's surface and the air above it that have been unaffected, or minimally affected, by humans; hence, the farther back we reach in time, the greater will be the extent of nature. Another way of putting it is this: Nature is what remains or what can recuperate over time when all humans and their works are removed.

These ideas of nature are a commonplace in today's world, thanks in part to their popularization in the environmental movement. They seem nonarbitrary, an accurate reflection of a common type of human experience and not just the fantasy of a particular people and time. But is this true? I believe it is. The nature/culture distinction, far from being an academic artifact, is recognized, though in variant forms, in all civilized societies—"civilized" itself being a self-conscious self-designation that postulates an opposite that is either raw and crude or pure and blissful. More generally, the distinction is present—in the subtext, if not the text—whenever and wherever humans have managed to create a material world of their own, even if this be no more than a rough clearing in which are located a few untidy fields and huts. I have already referred to the Lele in Africa. Their appreciation of a pure nature away from womenfolk, society, and culture is as romantic (and sexist) as that of modern American males. Thousands of miles away live the Gimi of New Guinea, another people of simple material means, whose bipolar *kore/dusa* is roughly equivalent to our "nature/culture." *Dusa* is the cultural and social world, opposed to *kore*, which means "wild"—the rain forest with forms of life, plant and animal, that occur spontaneously and hence are "pure."[18]

What about hunter-gatherers, who live off nature and have not carved a permanent cultural space from it? "Nature/culture" is unlikely to be a part of their vocabulary; they don't need it in their intimate, personal, and constant involvement with the individualized, all-encompassing natural elements. But since they undoubtedly feel at home in the midst of these elements, what an outsider calls wild and natural is to them not that at all; rather, it is a world acculturated by naming, storytelling, rituals, personal experience. This familiar world is bounded. Hunter-gatherers are aware that it ends somewhere—at this cliff or that river.[19] Beyond is the Unknown, which, however, is not "nature" as understood by other peoples. It is too underdefined, too far beyond language and experience, to be that.

A current trend in anthropological thinking is to wonder whether the nature/culture dichotomy is not more an eighteenth-century European invention than anything fundamental to human experience.[20] The binary in Western usage has fallen into disfavor because it is considered too categorical or abstract, and because it almost invariably sets up a rank order with women somehow ending at the bottom, whether they be identified with nature or with culture. One may also raise the linguistic conundrum of how far meanings must overlap to justify the use of European-language terms for non-European ones. I now offer one more reason for the declining popularity of the nature/culture binary. It is that one of the two terms has come to be dominant. In our time, culture seems to have taken over nature. Hardly any place on earth is without some human imprint. True, nature in the large sense includes the molten interior of the earth and the distant stars, and these we have not touched. But even they bear our mental imprint. Our minds have played over them; they are, as it were, our mental/cultural constructs. The ubiquity of culture in the life experience of modern people is surprisingly like that of hunter-gatherers, who, as I have indicated, live almost wholly in a cultural world with no nature, separate and equal, to act as a counterweight. There remains what I have called Nature with the capital *N*. But it, like the Unknown of hunter-gatherers, is beyond thought, words, and pictures. Whatever we touch and modify, whatever we see or even think about, falls into the cultural side of the ledger, leaving the other side devoid of content. Culture is, in this sense, everywhere. But far from feeling triumphant, modern men and women feel "orphaned." A reality that is merely a world ("world" derives from *wer* = man) can seem curiously unreal, even if that world is functional and harmonious, which is far from being always the case. The possibility that everywhere we look we see only our own faces is not reassuring; indeed, it is a symptom of madness. In order to feel real, sane, and anchored, we need nature as impact—the "bites and blows [of wind] upon my body . . . that feelingly persuade me what I am" (Shakespeare, *As You Like It* 2.1.8, 11); and we may even need nature as that which forever eludes the human mind.

But this hardly exhausts the twists and turns in meaning of nature/culture, real/imaginary. So the real is impact—the unassimilable and natural. But, as I have noted earlier, the opposite can seem more true. The real is the cultural. The cultural trumps the natural by appearing not so much humanmade as spiritual or divine. Thus, the cosmic city is more real than wilderness. The poem is more real than vague feeling. The ritual is more real than everyday life. In all of them there is a psychological factor that enhances the sense of the real and couples it with the divine, namely, lucidity. My own exposition of nature and culture, to the extent that it seems to me lucidly revelatory, is more real to me than whatever confused experiences I have of both. When I am thinking and writing well, I feel I have escaped to the real.

Escape to the Real and the Lucid

I began this chapter by noting that escapism has a somewhat negative meaning because of the common notion that what one escapes from is reality and what one escapes to is fantasy. People say, "I am fed up with snow and slush and the hassles of my job, so I am going to Hawaii." Hawaii here stands for paradise and hence the unreal. In place of Hawaii, one can substitute any number of other things: from a good book and the movies to a tastefully decorated shopping mall and Disneyland, from a spell in the suburbs or the countryside to a weekend at a first-rate hotel in Manhattan or Paris. In other societies and times, the escape might be to a storyteller's world, a communal feast, a village fair, a ritual.[21] What one escapes to is culture—not culture that has become daily life, not culture as a dense and inchoate environment and way of coping, but culture that exhibits lucidity, a quality that often comes out of a process of simplification. Lucidity, I maintain, is almost always desirable. About simplification, however, one can feel ambivalent. If, for example, a people's experience of a place or event is one of simplification, they may soon feel bored and dismiss it—in retrospect, if not at that time—as a thinly constructed fantasy of no lasting significance. Escape into it from time to time, though understandable, is suspect. If, however, their experience has more the feel of clarity than of simplification, they may well regard it as an encounter with the real. Escape into a good book is escape into the real, as the late French president François Mitterrand insisted. Participation in a ritual is participation in something serious and real; it is escape from the banality and opaqueness of life into an event that clarifies life and yet preserves a sense of mystery.[22]

To illustrate the wide acceptance of the idea that whatever is lucid feels real, consider two worlds of experience that superficially have nothing in common: academia and wild nature. Society at large has often called academia "an ivory tower," implying that life there is not quite real. Academics themselves see otherwise; it is their view that if they escape from certain entanglements of "real life," it is only so that they may better engage with the real, and it is this engagement with the real that makes what they do so deeply rewarding. And how do they engage with the real? The short answer is, Through processes and procedures of simplification that produce clarity and a quasi-aesthetic sense of having got the matter under study right. Now, consider wild nature. A sojourn in its midst may well be regarded as an escape into fantasy, far from the frustrations and shocks of social life. Yet nature lovers see otherwise. For them, the escape into nature is an escape into the real. One reason for this feeling certainly does not apply to academia. It is that the real *is* the natural, the fundament that has not been disturbed or covered up by human excrescences. What academia and nature share—perhaps the only outstanding characteristic that they share—is simplicity. Academic life is self-evidently a simpler organization than the greater society in which it is embedded. As for nature, in what sense is it simple or simpler—and simpler than what? However the answer is given, one thing is certain: People of urban background—and increasingly people are of such background—know little about plants and animals, soils and rock, even if they now live in exurbia or have a home in rural Idaho. Other than the few trained naturalists among them, their images of nature tend to be highly selective and schematic; indeed, for lack of both knowledge and experience, they may well carry reductionism further in the imaging of nature than in the imaging of social life, with the result that nature becomes the more clearly delineated of the two, more comprehensible, and therefore more real.

Middle Landscapes As Ideal and Real

Between the big artificial city at one extreme and wild nature at the other, humans have created "middle landscapes" that, at various times and in different parts of the world, have been acclaimed the model human habitat. They are, of course, all works of culture, but not conspicuously or arrogantly so. They show how humans can escape nature's rawness without moving so far from it as to appear to deny roots in the organic world.[23] The middle landscape also earns laurels because it can seem more real—more what life is or ought to be like—compared with the extremes of nature and city, both of which can seem unreal for contradictory reasons of thinness and inchoateness. Thinness occurs when nature is reduced to pretty image and city is reduced to geometric streets and high-rises; inchoateness occurs when nature and city have become a jungle, confused and disorienting. Historically, however, the middle landscape has its problems serving as ideal habitat. One problem is that it is not one, but many. Many kinds of landscape qualify as "middle"—for example, farmland, suburbia, garden city and garden, model town, and theme parks that emphasize the good life. They all distance themselves from wild nature and the big city but otherwise have different values. The second problem is that the middle landscape, whatever the kind, proves unstable. It reverts to nature, or, more often, it moves step by step toward the artifices of the city even as it strives to maintain its position in the middle.

Of the different kinds of middle landscape, the most important by far, economically, is the land given over to agriculture. People who live on and off the land are rooted in place. Peasant farmers all over the world—the mass of human population until well into the twentieth century—live, work, and die in the confines of their village and its adjoining fields. So the label "escapist" has the least application to them. Indeed, they and their way of life can so blend into nature that to visitors from the city they are nature—elements of a natural scene. That merging into nature is enhanced by another common perception of peasant life: its quality of "timelessness." Culture there is visibly a conservative force. To locals and outsiders alike, its past as a succession of goals, repeatedly met or—for lack of power—renounced, is lost to consciousness. Yet farmers, like everybody else, make improvements whenever they can and with whatever means they have. Their culture has taken cumulative steps forward, though these are normally too gradual to be noticed. Of course, one can find exceptions in the better endowed and politically more sophisticated parts of the world—Western Europe in the eighteenth century, for example. There, science in the broad sense of the systematic application of useful knowledge enabled agriculture to move from triumph to triumph in the next two hundred years, with far-ranging consequences, including one of psychological unease. An "unbearable lightness of being" was eventually to insinuate itself into the one area of human activity where people have felt—and many still feel—that they ought to be more bound than free. Nostalgia for traditional ways of making a living on the family farm is at least in part a wish to regain a sense of weight and necessity, of being subjected to demands of nature that allow little or no room for fanciful choice.

The garden is another middle landscape between wild nature and the city. Although the word evokes the natural, the garden itself is manifestly an artifact. In China one speaks of "building" a garden, whereas in Europe one may speak of "planting" a garden. The difference suggests that the Chinese, unlike Europeans, are more ready to admit the garden's artifactual character. Because artifice connotes civilization to the Chinese elite, it doesn't have quite the negative meaning it has for Europeans brought up on stories of prelapsarian Eden and on Romantic conceptions of nature. European gardens were originally planted to meet certain basic needs around the house: food, medicinal herbs, and suchlike. In early medieval times they were an indiscriminate mixture of the useful and the beautiful, as much horticulture as art. Progressively, however, the gardens of the potentates moved in the direction of aesthetics and architecture. From the sixteenth century onward, first in Renaissance Italy, then in Baroque France, gardens were proudly built to project an air of power and artifice. The technical prowess that made playful fountains and mechanical animals possible, together with the garden's traditional link to the phantasms of theater, resulted in the creation of an illusionary world remote indeed from its humble beginnings close to the soil and livelihood.[24]

A striking example of the pleasure garden in our century is the Disney theme park—a unique American creation that, thanks to modern technology, is able to produce wonder and illusion far beyond that which could be achieved in earlier times. Unique too is the theme park's erasure of the present in favor of not only a mythic past but also a starry future—in favor, moreover, of a frankly designed Fantasyland peopled by characters from fairy tales and from Disney's own fertile imagination. What is more escapist than that? In the spectrum of middle landscapes, a countryside of villages and fields stands at the opposite pole to a Disney park. The one lies closest to nature; the other is as far removed from it as possible without becoming "city."[25] Disney's carefully designed and controlled world has often been criticized for encouraging a childish and irresponsible frame of mind. But again my question is, What if culture *is*, in a fundamental sense, a mechanism of escape? To see culture as escape or escapism is to share a disposition common to all who have had some experience in exercising power—a disposition that is unwilling to accept "what is the case" (reality) when it seems to them unjust or too severely constraining. Of course, their efforts at escaping, whether purely in imagination or by taking tangible steps, may fail—may end in disaster for themselves, for other people, for nature. The human species uniquely confronts the dilemma of a powerful imagination that, while it makes escape to a better life possible, also makes possible lies and deception, solipsistic fantasy, madness, unspeakable cruelty, violence, and destructiveness—evil.

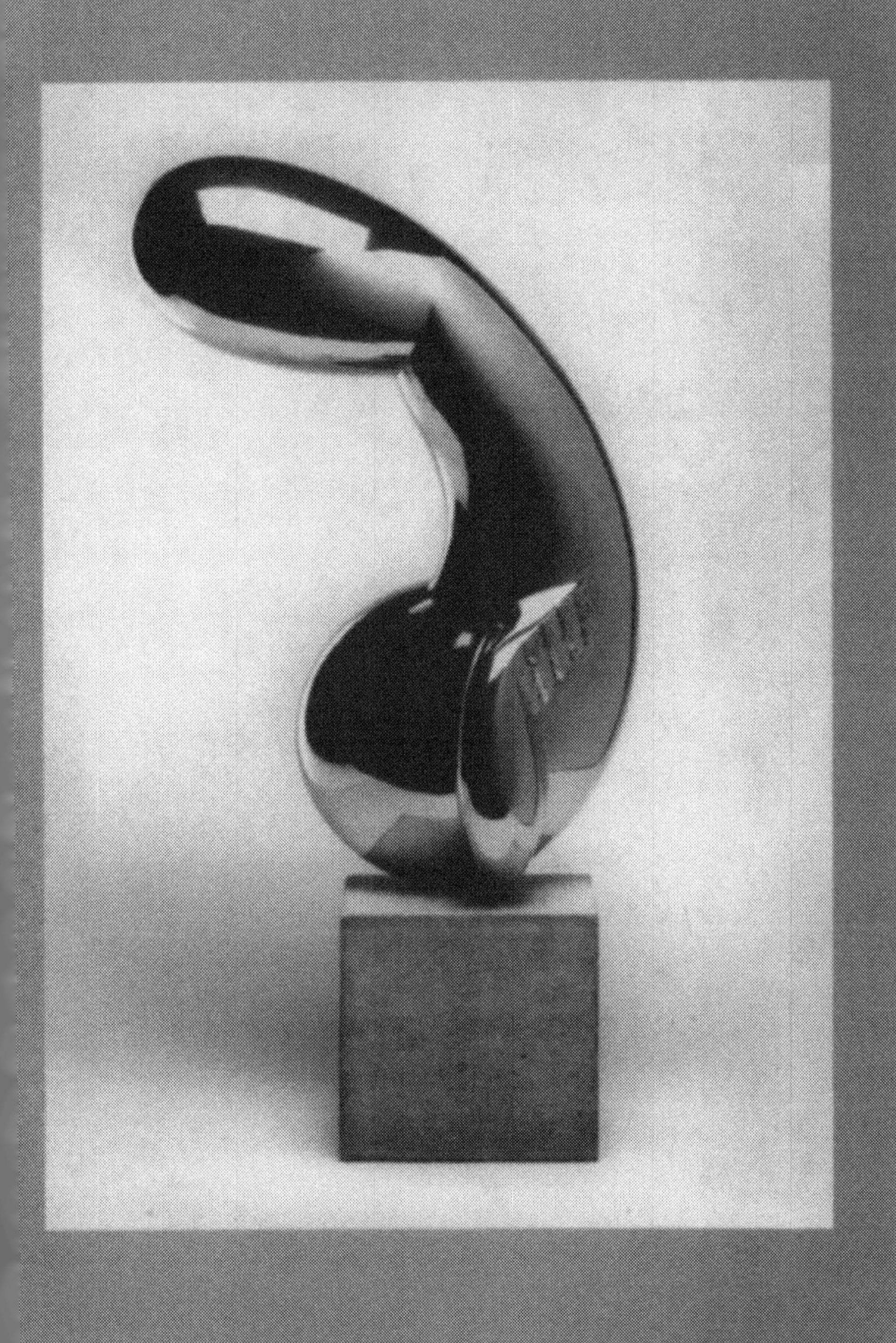

2

ANIMALITY

Its Covers and Transcendence

• • •

I have broadened the concept "escape" so that it means both literal flight and steps taken to change or mask an unsatisfactory condition—scrubland, urban neighborhood, whatever. Any physical place, natural or artifactual, can be altered for the better; people do it all the time. But what if the "place" one wishes to escape from is one's own body? Can one take flight from one's own corporeal wrapping? The very idea is science fictional. Yet we do it often—imaginatively, as in daydreaming, or when we become wholly absorbed in some other person, object, or event. If one can't escape literally from an unsatisfactory body, can one rebuild it? Well, there is plastic surgery, and some people in the entertainment industry have taken rather extreme steps to alter their appearance.[1] As for internal organs, when they malfunction they can, in many instances, be repaired and even replaced. But these remain minor adjustments.

At an everyday social level, dissatisfaction is not with the body as such, but rather with its imperfections as measured by the standards of the time. At a deeper level, dissatisfaction may be directed at the body simply because it is an *animal* body, with offensive odors and fluids, humors and imperious biological demands.

No matter how comely in its own way, the body can still be an embarrassment, a reminder that our reach toward some more elevated status called "human" or "spiritual" is tenuous—hostage to the sewer odor of flatus.[7] Paradoxically, this very awareness of being an animal makes us human, for it is a problematic awareness unknown to other animals. Self-awareness is an ambivalent gift. Endowed with the ability to take a position outside myself, I can look back on this self and see it in all its corporeal grossness. But why do so when the result is mortification? More bolstering to self-esteem is to turn my gaze on some other man's or woman's animal self. Or better still, more elevating morally, is to turn my gaze on the cultural world. The first two turns of the gaze drive me to the third. Embarrassed by animality—my own as well as other people's—I and my fellow human conspirators cover it up, escape into a world of cultural creations that reassures us of our exceptionalism.

The story of cover/escape shows an overall progression from less to more elaborate forms of art and artifice. But that progression is not straightforwardly linear. It has twists and turns because that is how our consciousness and feelings work. The value they aspire to may turn at some point, unexpectedly, into its opposite, such that, for example, the animal rather than the human (cultural) state is preferred. To complicate matters, wanting to be more natural and animal-like is itself a cultural attainment. "Less is more," and even as one strips down, one may become more rather than less sophisticated. Another point to bear in mind is this: It has often been said that humans are animals that possess culture, which amounts to saying that they possess both the desire and the ability to escape from or cover up their condition. Wanting to reach a better condition and having the cultural skills to do so, being embedded in the very definition of humanness, cannot therefore be a deviation, and it certainly is not deviance. Yet in many societies it was and is often seen as such. Not only are "cover" and "escape" strangely ambivalent words; even "transcend" can carry a negative meaning of arrogance or excess. Still, overcoming animality is a universal human undertaking, one in which some people have shown greater ambition or have been more successful than others. I shall tell this story of overcoming by exploring three prominent indices of our animality: food and eating, sex and procreation, and dying and death.

Food and Eating

Eating is both necessity and pleasure—until we pause to reflect on what we are doing, at which point a feeling of unease obtrudes. Is this uneasiness widely shared among humans, a hidden sensitivity that can be aroused under the right circumstance? What would be a right circumstance? Perhaps the most common is when we watch others eat or are watched as we eat. Eating may be a joyous public occasion in which many people participate, it may be accompanied by much ceremony, but it is not itself a public performance. Eating, humans realize, is animal and must remain essentially private, a condition they ensure by creating a space for it and protecting it against the appraising eye. Animals at the zoo cannot have such privacy; indeed, they are assumed to need none. One of the more popular events at the zoo is the feeding time of the great carnivores. People feel a combination of awe and superiority as they watch a lion chomping on raw meat. To human eyes, the lion then is no longer king, a regal figure with its crown of mane, but simply a feeding animal, one that, moreover, did not obtain its own food but was given it; and that burdens the animal with an extra measure of humiliation, which people enjoy watching.

Ethnographic Literature As Watching

One reward of reading the older ethnographies in today's austere moral climate of correctness is that it provides us with the guilty pleasure of watching. We "watch" as we read about others, their habits and customs. An anthropologist of the 1920s tells us in the superior tone natural to his time that on Diomede Island the Eskimo hunters

> would squat around their meal of long-buried walrus meat which lay on bare ground strewn with the excreta of dogs and people. The raw, rotten meat was reminiscent of highly ripe cheese. . . . [Bering Strait] Eskimos bury fish heads and allow them to decay until the bones become the same consistency as the flesh. Then they knead the reeking mass into a paste and eat it. A few other examples of Eskimo food will indicate his lack of squeamishness: raw intestines of birds and fish, swallowed like oysters; live fish gulped down whole, head first; slime scraped from a walrus hide together with some of the human urine used in the tanning process . . . fat, maggoty larvae of the caribou fly, served raw; the contents of the caribou's paunch, left in the body so long that the whole mass has become tainted; deer droppings, munched like berries, or feces taken from the rectum of this animal.[3]

Did the Eskimos like being watched while they ate? I doubt they did, for it must have made them feel self-conscious. The anthropologist himself got a certain amount of pleasure out of observing, as we can judge from his prose. And what about us who read him?

Reading anthropologist Colin Turnbull's account of the Mbuti Pygmies of the Congo rain forest provides much innocent pleasure, for here is a people who until recently seem to have lived in Eden, a benign natural environment that is the polar opposite of the Eskimos', supporting a way of life in which there is no recognition of evil. It comes as a shock, therefore, to read in another writer's book that this Eden too is stained by the sort of bloodiness inevitable whenever big-game hunting occurs. The picture that emerges in my mind is a disturbing mix of skill and courage on the one hand and on the other hand the unique human power, even in a group deemed "primitive," to rain havoc (temporarily) on nature. Wielding only a spear, a Pygmy hunter can bring down an elephant, thus demonstrating admirable courage and skill. What follows is less inspiring and is certainly not Edenic. Tropical heat triggers rapid decay. The beast's belly starts to swell. A hunter climbs up it and dances in triumph. Someone plunges a knife into the belly, letting out an explosion of foul-smelling liquid and gas. Pygmies swarm over the carcass in an orgy of butchering. Within a couple of days a patch of forest becomes a blood-spattered field of skin, bone, and intestines. Celebration may continue for weeks. People gorge themselves with meat and sing and dance with mounting erotic fervor.[4]

A British travel writer recently observed, with a mixture of revulsion and satisfaction, a guzzling feast in Canton, China. As readers of his seductive prose, we share his satisfaction, his sense of superiority. Spread before us is the following scene. A dozen men—small traders and merchants—gather at a table in a floridly decorated restaurant. Their expectation is unbearably tense: "Every course drops into a gloating circumference of famished stares and rapt cries. Diners burp and smack their lips in hoggish celebration. Bones are spat out in summary showers. Noodles disappear with a sybaritic slurping, and rice-bowls ascend to ravenously distended lips until their contents have been shovelled in with a lightning twirl of chopsticks."[5]

The Prestige of Meat

Reading about the eating habits of people in other places or earlier times reminds us of how until a certain stability was achieved by modern means of food production and distribution, human lives—outside a few favored areas, including prosperous cities—alternated between feast and famine. In the rare times of plenty, almost everything that could be eaten was eaten, and in astonishing amounts. The menus of the past surprise us by their inclusiveness. Yet this inclusiveness ought not to surprise us, for one reason why humans have spread successfully into so many geographical environments—one reason why they have become "lords of the earth"—is their ability to chew and digest almost anything. Most societies nevertheless have some kind of food taboo, and some have strict dietary laws, one purpose of which is to distinguish members of that society from others and to allow them to feel superior (less animal, more spiritual). But before I describe the various attempts at transcendence, I need to address an apparent contradiction in my thesis. If gorging food is a sign of animality, if eating meat is the clearest exhibition of carnivorous brutishness, then why have the powerful and the rich in the Western world stuffed themselves openly in the past, with no sense of shame, and why has the consumption of meat, not only in the West but in other parts of the world as well, become a signature of status?

Consider the ancient Romans, the medieval Europeans, and their descendants. Civilized in so many ways, the Romans were remarkably uncouth in their feeding habits. We see them lounging around a table groaning with food, their fingers sticky from gravy-drenched meats, for they had few utensils; foods were wrapped in pastry so that they could be lifted cleanly to the mouth, and additional shells were provided to scoop up the juicier slippery bits.[6] Eating manners hardly improved in medieval and Renaissance Europe, even among powerful people who had no need to stuff themselves in anticipation of lean times ahead. Indeed, manners did not show real finesse until the eighteenth century. Movie scenes that depict England's Henry VIII chomping on a leg of mutton held up by a bejeweled hand are probably fairly accurate. Here is a real feasting Lion King! If we who watch the film feel superior to uncouth Henry, such could not have been further from the thoughts of participants in the lively event itself—the courtiers, pages, and servitors who if anything felt only admiration and awe.

The sheer quantity of food made for prestige, but what really counted was the amount of animal flesh. At a Twelfth Night feast in mid-seventeenth-century England, *each* guest was expected to guzzle his way through seven to eight pounds of beef, mutton, and veal.[7] Meat remained supreme in the West's hierarchy of gustatory values till the last quarter of the twentieth century, when, increasingly for health reasons, it lost its former eminence. Yet it retains a certain aura. As one hostess notes, when she serves food to her guests it still seems natural to put the roast beef on the plate first, then the vegetables; first the principal actor makes an entrance to admiring "oohs" and "ahs," then come the supporting players.[8] This bias of the West is, however, far from unique. Perhaps the carnivore in the human beast has something to do with it. Most people, historically and in different cultures, simply like cooked meat. What makes one sniff the air in anticipation? The fragrance of a roast, not that of cabbage. Texture too matters. Think of the meat's chewy resistance to the teeth and its subsequent collapse into a rich paste that coats the cavern of the mouth. Even when a religious culture turns to the ideal of eating only vegetables, as in Buddhism, artists of the palate strive to give them the flavor and texture of meat. The nutrient value of animal protein no doubt plays an additional role. Eating meat seems to make people feel stronger, more packed with substance.

Apart from these intrinsic qualities, there is the external factor of prestige associated with the hunt. Recent research among hunting-gathering bands and among people whose economy combines agriculture with hunting shows that meat, rather than the plant foods that make up the bulk of the diet, is the delicacy. Gender roles come into play here. Women forage for edible plants and trap small animals in the camp's neighborhood, to which they are confined by the needs of nursing infants and young children. Where a people also practice agriculture, the women may have primary responsibility for taking care of certain crops and vegetables grown around the village. In one way or another, they contribute most of the food, but their means of obtaining it are dull and routine. By contrast, the men hunt, a far more exciting and glamorous activity. Hunting requires that the men move out into a larger and less known world. The fact that this world can be dangerous calls for close cooperation and bonding, which are gratifying in themselves. Moreover, unlike women's work in the village fields, hunting is a story with a plot climaxing in a dramatic kill, a story the telling of which back home yields extra prestige. To put it a little differently, hunting is an exhilarating escape from the constraints of rooted life. Men in preliterate communities appreciated it, as many men in modern society still do. Game meat too tends to be more highly regarded than the meat of livestock, as though the freedom of wild animals to roam gave their flesh a virtue that confined domesticated animals do not and cannot have.[9]

Masks and Distractions

Food is life—energy and power. Having enough to feed not only oneself but others less capable or fortunate boosts the provider's ego. All this is well known. Food's preeminent place in life is hardly a puzzle. At the same time, however, shoveling it down one's gullet is stark animal behavior—one end of a process the other end of which is the evacuation of urine and feces. Coarse eating may not arouse universal disgust, and table manners may be so elementary among some people as to seem nonexistent. Nevertheless, an underlying judgment of food guzzling is implied, if only in the sense that no culture has raised it to the status of commendable practice.

Among masking devices, the most common is to turn eating into a social rite. People chewing food can pretend that they are engaged in loftily listening to their neighbor's remarks, the music of a flutist, or even poetry read aloud by an educated slave—a practice of proper Romans in ancient times that somewhat excused their sloppy eating. To take a more modern example, Dr. Samuel Johnson was a notoriously coarse eater. On one occasion he demanded the boat containing leftover lobster sauce and proceeded, to the horror of his fellow diners, to pour the thickening goo over his plum pudding. If he remained an acceptable, indeed popular, guest at the dinner table, it was because he had the wisdom and raconteurial skills to compensate.[10]

But if eating is to become a genuine social event and ceremony, manners must be refined. In Europe from the late Middle Ages onward, aspiring courtiers and the upper class took steps in that direction. They adopted new utensils that put a distance between the food and themselves—outstandingly, the fork (a metal claw), which substituted for the human hand in performing unseemly tasks. Old utensils were refined—for example, the knife, the sharp tip of which was rounded off in the sixteenth century so that it would seem less an instrument of violence. The more elegant eaters stopped spitting bones onto the floor, and they learned to masticate with the mouth closed, noiselessly.

The animal origin of the food itself could be masked as well. From 1700 onward in Europe and particularly on the Continent, meat was less frequently served whole. Elizabethan England even enjoyed a vegetable renaissance.[11] English gardeners had long recognized the medicinal and hygienic virtue of herbs, but in the sixteenth century they extended the property to a number of vegetable foods, encouraging their greater use. A dent was thus made in the picture of man as a natural carnivore. Delicacy of taste, aroma, and texture increasingly came to matter more than the sheer bulk or cost of the food. Cooking strove to become architecture and art, and as such it successfully hid from consumers the fact that the materials used had once been living animals. Language might also be enlisted as a masking device. What are we eating? Not cattle, pig, or deer, but beef, pork, and venison, words of non-English origin (think of the menus of fancy restaurants, where what we eat is buried under a flourish of incomprehensible foreign words). As for plants, it is significant that we never consider picked fruit and cut vegetables as mutilated and dead. Our terms of evaluation are "fresh" and "stale" rather than "alive" and "dead," though the terms "rot" and "rotten" are used.

Societies differ in their sensitivity to the animality of eating. The Chinese elites have been, historically, among the more sensitive. Since the earliest dynasties they have striven to cover up the grosser aspects of food consumption. An ancient practice that has become a signal trait of Chinese cooking is the shredding of meats and vegetables into small pieces that are then variously recombined. The origins of the materials are thus hidden. But perhaps at a conscious level what the Chinese chef desires is not so much to hide something as to experiment with new flavors and textures, to elevate cooking to an art—and indeed to something beyond art, for art in ancient China was also cosmic world-view, ritual and religion. Food and eating thus found themselves propelled into exalted realms.

Consider health. An overriding idea among the Chinese is that the kind and amount of food consumed are intimately related to health; food is also medicine. Health is ultimately an effect of cosmic harmony, and it was and is the Chinese view that foods play there a mediating role. They can do so because they are believed to have either yin or yang qualities, yin and yang being the two principles of the cosmos. As early as the Chou period (ca. 1100–220 B.C.) illness was considered a disequilibrium of yin and yang forces in the body, which could be set right by foods of one kind or another.[12] Food and eating are an essential part of the cosmic ritual. *Li Chi*—the Book of Rites, a Confucian classic with material dating back to the fifth century B.C.—is full of references to the right kinds of food for various occasions. It is a work of elaborate etiquette, except that etiquette here is a system of reverend attitude and reverend gesture that goes far beyond mere social manners.[13]

Food enters the Chinese understanding of health, ethics, and religion in other ways as well. One ideal of food is its freshness and natural flavor. As men and women should be what they seem—open and pure, honorable and trustworthy—so food should be pure and fresh, its natural flavor not hidden under a meretricious covering of sauces and spices. Taking the parallel one step further, the Chinese believe that just as such qualities in human beings promote social harmony, so in foods they promote the harmonious functioning of the internal organs—that is, individual health.[14] Another ancient Chinese ideal is "the middle way"—nothing to excess. Excesses of crudity—the frenzy of eating in the Cantonese restaurant noted earlier—are to be avoided, but also excesses of refinement that can tempt a sophisticated society. Must meat be cut and presented in a certain way before one would deign to eat it? Confucius himself, though a strong advocate of "the middle way," seems to have at times gone pretty far in that regard.[15] Must tea be sipped only from the finest china? Does one lose face if one can't distinguish rainwater from that which comes from a stalactite cavern? Perhaps; but a man of real taste knows how to distance himself from extreme refinement. The great playwright of the thirteenth century, Kuan Han-ch'ing, included in one of his dramas the following lines of rural rhapsody: "The fall harvest is gathered in. Let's set a feast under the gourd trellis, drink wine from earthen bowls and porcelain pots, swallow the tender eggplants with their skins, gulp down the little melons, seeds and all."[16]

Gender Difference

One way to indulge the animal appetite and still feel superior is to make the unabashed claim of being a "king among beasts." Raw animal power itself then commands admiration and deference. I have already broached this point, but it now calls for a further comment along gender lines. Consider the West, where men have not hesitated to show, even boast of, their kinship with carnivores. In the *Iliad* Homer repeatedly likened heroes to wild beasts—eaters of raw flesh. Germanic parallels suggest that ancient Indo-Europeans were in the habit of thus portraying heroic behavior.[17] When rising above one's lustful and sinful self became the ideal, as among the religious in medieval Christendom, men and women showed surprisingly different ascetic styles. A monk would have struggled to abstain from sex; should he succumb to temptation, the fact would have been acknowledged only in the confessional. But a Thomas Aquinas, no doubt abstemious sexually, could openly enjoy his food and be proudly wide of girth. A religious woman, by contrast, had to be both virgin and sparing in diet. Far more than an obese abbot, an obese abbess presented a troubling image. Gender differences were equally manifest in secular Europe and its offshoots overseas. Elizabeth I might have been more of a man than her effete courtiers, but it is hard to imagine her biting into a leg of mutton, as one so readily can her father, Henry VIII. In the frontier towns of America, women stood for refinement and culture, men and boys for nature, and one way for the males to demonstrate their uninhibited naturalness was to wolf down their food. In Victorian England, men could eat heartily, provided they knew how to distinguish the fish knife from the butter knife. Women showed their even greater distance from biology by eating little, and that with as much daintiness as they could muster. Putting food into the mouth was itself animal enough; actually eating might even make one fart—a truly devastating revelation of animal being.[18] Even in contemporary modern society, men seem to feel that one way to exhibit their maleness is pointedly to ignore the finer points of etiquette. The "Animal House" of movie notoriety can only have been a fraternity.

Of course, I have been highly selective; a hearty naturalness among males has by no means always been considered desirable. Thus, from the late sixteenth to the eighteenth century, European men of high status, or with high social aspirations, took pride in being dainty and foppish. Perfumed, coifed, and teetering on high heels, the men were in every way a match for their ladies in elegance. Hunting culture and a warrior ethos—the two often overlapped or fused—promoted gender difference. They did so in Europe until early modern times—indeed, as we have seen, in our own time. They also did so in medieval Japan, with its stereotypically silent, gruff males and porcelain-delicate females. China was a notable exception, for despite being an empire, it lacked a warrior class. Near the top of its social hierarchy was the scholar-official, his weapon the brush. Warriors commanded admiration only if they were also clever strategists and showed a literary flair. The expression "pale-faced scholar" was complimentary rather than, as it would be in the West, a put-down.[19]

What has happened? What has become of my thesis that escapism is a human universal? My answer is that even in societies where the hunter-warrior ethos dominates or lingers, the men are as committed to escape as are the women; they differ only in the means. Men choose to boast of rather than to hide their animal traits. This choosing to "act the noble beast" is a variant form of "returning to nature"—a move prompted by the need to escape from society's debilitating and unmanning elegancies.

The Prestige of Abstinence

One route to prestige is to act the powerful animal. Another and opposite route is to make the claim of being a cultivated or spiritual person, so far removed from animality that one hardly eats. History is rich in examples of people who deny themselves food in answer to some higher calling, or simply to appear superior. This denial of appetite and the animal self is not exclusively a religious affectation of high culture. It was manifest, for instance, among people of the soil and of livestock in Rwanda, a central African country to which the world paid little attention until 1994, when slaughter of genocidal ferocity broke out between its two dominant ethnic groups.[20] As Jacques Maquet has shown, one source of the deep-seated animosity between them was the racist notion of "inferior" and "superior," the one designating bondage to earth and the animal estate, the other implying an ability to rise triumphantly above both.[21]

Rwanda's population was made up mostly of Hutu farmers, a sizable minority of Tutsi pastoralists, and a small minority of Twa hunters, who supplemented their income by singing, dancing, and clowning. A strongly differentiated social hierarchy began to emerge during the late nineteenth and early twentieth centuries, based on racial stereotyping and biases that already existed among native populations but were inflamed by the country's European overlords. The Tutsi were encouraged to see themselves as an aristocracy, tall and of slender build, and to see the Hutu as being shorter and stumpier, with the additional unflattering traits of woolly hair, broad flat nose, and thick lips; in time the oppressed Hutu came to see themselves that way as well. As for the Twa, both the Tutsi and the Hutu regarded them, half jokingly, as closer to monkeys than to human beings.

Food habits further differentiated and distanced Rwanda's ethnic groups. The Tutsi diet drew mainly on dairy products and tended to be more liquid than solid. A meal usually consisted of cooked slices of sweet bananas, a bread made with sorghum flour, and lots of milk. When they ate meat, which was only rarely, it was cut into small pieces and boiled. The Hutu ate more, though their food was less refined. Popular among them was a kind of porridge with beans, peas, or maize. The sweet potatoes they consumed in large quantities were considered by the Tutsi too common to eat. The Twa liked to eat and drink as much as possible whenever they could, for their food supply was irregular, being dependent on a successful hunt or a generous reward for dancing and singing. Conceivably these dietary differences might have remained just that. In fact, they were integrated into conceptions of superiority and inferiority. The Tutsi behaved as if the need for nourishment was beneath their dignity. Eating should be done in private, they believed. Friends would be offered beer or milk, but they would not be asked to share a whole meal. Some older Tutsi took pride in subsisting only on liquids. When the Tutsi went on a journey of less than three days, they might not eat anything at all. As pastoralists, they liked to behave as though they did not depend on the foodstuffs produced by the Hutu. Eating less and differently was a way of elevating themselves above their fellows. The Tutsi have legends telling them that they come from another world—that they are human, but not in the same way as the Hutu, who are greedy, or the Twa, who are gluttons.

In an ideal world, people may need nourishment, but their food is not made up of animal meat. Adam and Eve were vegetarians before the Fall. In the Taoist paradise, people didn't even have to eat plants; when they felt tired and hungry, they would drink the water in the rivers and find their vitality restored. A much later account of an unspoiled world, written during the T'ang dynasty, presents a people of childlike exuberance who lived presumably on fish but not, heaven forbid, four-footed mammals: "They swarm to the tops of trees, and run to the water to catch bream and trout."[22] The Olympian gods were vegetarians, feeding gloriously on ambrosia and nectar. Why couldn't humans do the same? In the mythic golden age of ancient Greece, people did not have to maim and kill to survive, nor did they have to stuff themselves, for nature provided food in abundance. They were free to practice moderation, a supreme Greek virtue. Killing was violence, an egregious offense against moderation, and against peace. Pastoralism and farming were successive falls from the golden age. Pastoralism, though more benign than farming, still led people to desire property, from which emerged a spirit of contentiousness. Agriculture aggravated that desire and created conditions that encouraged war. Domesticating and killing animals and perhaps eating their meat fueled an innate human capacity for violence.

The ancient Greeks had several arguments for refraining from meat. One was that animals, like people, had "souls," and that there was a time when animals and people shared a common language. The quasi equality thus postulated made the eating of animal flesh unconscionable. To escape the intolerable conflict this belief entailed, the Greeks qualified it, saying that not all animals were besouled. They further distanced themselves from animals by claiming that man's greater intelligence enabled him to converse with the gods in a supernal realm. However, for that to happen, he must first free himself from bondage to his animal body. In Greek thought, the gap between plants and animals, animals and men, body and spirit, tended to broaden over time. Already evident in Pythagoras, the theme became even more pronounced after his death in the writings of his followers. The elevation of spirit over body reached its peak in Plato, whose unrelenting hold on Western thought and feeling, both religious and secular, persisted until our own populist, antihierarchical age. If Plato advocated vegetarianism for philosophers, it was less from a sense of fraternal bond with animals than from a feeling of revulsion toward the lower appetites, tethered to the stomach "like a beast untamed."[23]

In the Middle Ages, Christianity favored periodic fasting for a variety of reasons, including health, penance, and propitiation, and even as a quasi-magical way of inducing fertility, based on the idea that in the natural world scarcity was almost always followed by abundance. Predictably, ascetic Christians also sought to discipline the body so that the spiritual-angelic self could emerge into the purity of light. Patristic writers "cited not only the Old Testament models of David, Esther, and Judith, who by fasting offered pure hearts to God," but also classical writers, including Pythagoras. Clement of Alexandria (d. ca. 215) expressed the view that "fasting empties the soul of matter and makes it, with the body, clear and light for the reception of divine truth." In the homely formulation of the abbot Nilus (d. 430), fasting helped one to pray, whereas a full stomach made for drowsiness. In the seventh century, Isidore of Seville summarized much of the patristic tradition when he spoke of fasting as "the doorway to the kingdom, the form of the future, for he who carries it out in a holy way is united to God, exiled from the world, made spiritual."[24]

Eating and Spirituality

Despite all that has been said about the spirituality of fasting, one may still ask, Cannot eating also contribute to spiritual elevation? Christians in particular must confront this question, for their most sacred rite is the Lord's Supper (Holy Communion). Christ's last supper with his disciples both enforces and contradicts certain common assumptions about a shared meal. On the one hand, it dramatizes the idea that eating at the same table strengthens human bonding; on the other hand, betrayal is shown to occur in its midst. The betrayal is not just incidental to the Last Supper; it is crucial to it, for without the betrayal how can there be the agony in the garden and, ultimately, human salvation? That so much good can come out of evil is a profound paradox in Christian theology, but perhaps the story of the Last Supper resonates for another, nontheological reason, one that rests on human experience, namely, that the very power of the shared meal to cement friendship provides the opportunity for its betrayal.

The Last Supper has another message, which, if understood correctly, would have radically changed the meaning of eating meat in the Christianized West. How so? Consider what happens at the Last Supper. Bread and wine, unproblematic foods derived from plants, are consumed. But vegetarianism is hardly the issue. The Gospels do not bring the story of the Last Supper forward as a solution to the moral ambiguity of eating. Quite the contrary, for Christ shockingly tells his disciples that the bread and the wine are his own body and blood. Eating and drinking are thus plunged back into their luridly animal nature. Predictably, the more acerbic critics of Christianity have accused its followers of being cannibalistically inclined. In fact, what takes place in the Last Supper is the opposite of incarnadine animality. Eating on that occasion, and on all the subsequent occasions that commemorate it, is so spiritualized that it loses almost all connection with the consumption of animals and plants. Eating is become a sacred ritual at which one human being (following Christ's example) freely gives of himself ("even unto death"), and at which another accepts the gift, with the understanding that he in turn may have to offer his body and blood as food.

Destruction Precedes Construction

In Eden or heaven, eating shouldn't be necessary, since teeth grinding down even plant food is patent violence. On earth, however, destructing and incorporating another organism is how any animal assimilates energy and maintains its body. There is no other way. In the cultural realm too, destruction generally precedes construction. Something has to be "chewed up" before other things can be put up. An outstanding exception is storytelling, which can be practiced without having to destroy something first. Other fine arts, from weaving to sculpture, do require some degree of prior destruction. Visit a professional craftsman's workshop; it can seem a slaughterhouse, except that here the broken and torn viscera of nature—pieces of marble, stone, and wood, bent metal and filings—are mostly inanimate.[25] If, generally speaking, even the crafts and fine arts require some degree of violence, the scope of the destruction involved in the raising of humankind's larger cultural works—farms, towns, and cities—boggles the mind; more, it can ignite feelings of moral revulsion. Some people, including ecologists of our day, have wished a return to simpler ways of life. But how far back should human beings go? And how simple must a way of life be before it can be considered innocent? The dilemma in its extreme form revolves around the human body, which too is a construction, built up with plant and animal parts.

The covers of culture over the animality of eating are so successful that we—that is, people generally—do not give it thought. Eating? When so reminded, we may say with a touch of irritation, "Of course, we eat. We are, after all, animals in the food chain." Moreover, contrary to my earlier claim, we not only enjoy eating, we can also enjoy watching others eat. What proud cook doesn't delight in his or her guests' ravenous appetite? Few satisfactions are greater in a mother's life than watching her child gobbling down food. Yet it remains the case that once we have peeled off the cultural covers to ponder the moral implications of this inexorable and most elemental form of violence, it is hard to retain our innocence. A dilemma infects the core of our daily existence from which there is no escape.

Sex and Procreation

As animals, we eat and we copulate. But where human beings are concerned, there is a world of difference between the two. Imagination applied to food has produced some great dinners—has induced sensual satisfactions of a very high order, but hardly anything that touches the soul. Not the worst of vices, gluttony is nevertheless the most contemptible, says Allan Bloom, for it is "a sign of small-mindedness. One can fantasize about a great dinner, but if that's the limit of one's fantasy life, it is a puny thing, morally and aesthetically handicapped." By contrast, "sex, which is in fact no more naturally spiritual than eating . . . is able to produce the most splendid flights of the soul as well as terrible tragedies."[26]

Eating and sexual intercourse are both necessary to the continuance of life. Both entail violence, but in the one it causes the evisceration of plants and animals while in the other it is more a playful roughhouse of thrusts and bites, an ecstatic frenzy of love, giving intense pleasure to the entwined couple without damaging, in the process, anyone else. In eating, a bloated feeling must sooner or later be relieved by the expulsion of odorous waste; in contrast, the postcoital exhaustion that follows orgasmic union is a reminder that, for all the sense of fulfillment, effort has been expended, some vitality given up, the result of which is the possible birth of a new life. Eating, even in the presence of others, is a replenishing and pleasuring of self; that is what eating is, by definition. Sex, on the other hand, is engagement with another; even in masturbatory fantasy, the image of another must be conjured. Can these be some of the reasons why sex has spiritual dimensions absent from eating? Yet whereas unmannerly eating is deemed merely gross, the display of tumescent sex other than to a lover in secluded privacy is widely considered obscene.

Voyeurism and Obscenity

Sex is the ultimate private act. Ethnographers who do not hesitate to intrude upon the most intimate behaviors in the name of their science hesitate to observe and record the sex act. I noted earlier that feeding times are popular at the zoo; we delight in seeing lions chomping on raw meat in uninhibited animality. I would now add that monkey cages are also popular; they seem to exert a particular fascination for adolescents, yet who other than infants can claim total indifference? Here are our primate cousins, similar to us in so many ways, yet radically alien in their unselfconscious sexual exploration of each other.

Is it really possible to watch chimpanzees in estrus in cool scientific detachment? The mating instinct and its physiological manifestations, which humans find embarrassingly and inconveniently forceful in themselves, seem inordinately exaggerated in nonhuman primates. Human tumescence is chaste indeed compared with the flamboyant genital swelling of the female chimpanzee in the peak stage of her readiness for male mounting. Large, protuberant, and pink, chimpanzee genitalia advertise with a blatancy found among humans only in their pornographic art. As for prose, the explicit accounts of sexual congress that are pornography's specialty appear matter-of-factly in animal-behavior literature. But is it really possible to read them matter-of-factly? What exactly does the reader see and feel upon encountering a passage such as the following? "During a reunion with a female [chimpanzee], a male often inspects her genital area. He may bend close and smell her bottom directly, or poke his finger into the vulva and then sniff the end of it . . . two or three times." This poking inspection of the female's sexual organ in disregard of her own sense of self would be judged obscene if the players were human. And so I am led to the impolite question, Can the ethologist focusing his binoculars on the performance, or the reader who sees it all in the mind's eye, altogether escape opprobrium? Is science a cover?[27]

As the female chimpanzee advertises aggressively, so does the male, drawing attention to his breeding potential by various means—bipedal swaggering, the shaking of a branch, and suchlike—but most directly by exposing an erect and bright pink penis, normally flaccid and concealed in the prepuce, against the white skin of the thighs and the lower abdomen. Compared with such uninhibited advertising, the sex act itself is surprisingly subdued, by human standards. In fact, there is little bodily contact as such, no tight locking of two bodies into one. Rather, the male merely leans toward the female's back as he makes his insertion.

Touching, other than the insertion, may be little more than the male grasping his mate with one hand to steady himself in his squatting position. Orgasm there may be (primatologists are not yet sure); certainly the chimpanzees pant and squeal in excitement, but there is nothing like "divine madness" in the coupling—no sweaty overpowering passion that expels the world.[28] The act can indeed seem a merely necessary function, a practical step accompanied by a certain amount of pleasure, in obedience to nature's procreative goal. So in the end, the act, for all its showy physiological billing, is a disappointment to the prurient voyeur, as it is to the prurient reader of the ethological report.

Reading is voyeurism; pictures emerge from the printed page as one reads. The words are put there by a writer, who encounters a different kind of challenge when he describes eating than when he describes sex. Eating can be presented neutrally, if that is his wish, but sex cannot. Whereas ordinary words such as "hand," "nose," and "shoulder" designate body parts without strong emotional coloring, other parts resist plain utterance. The sex vocabulary available to the middle class is limited to four types, according to C. S. Lewis: "a nursery word, an archaism, a word from the gutter, or a scientific word." Hard as one tries at mere description, one produces willy-nilly baby talk, arch speech, coarseness, or technical jargon. Certain common words have become "obscene" because they have long been "consecrated (or desecrated) to insult, derision, and buffoonery." To speak them at all is to evoke the randy-machismo atmosphere of the slum, the barracks room, and the all-male school.[29] It is the word that is offensive, rather than the anatomical feature to which it is applied. It offends by reducing a human being to a single organ, the (often involuntary) arousal of which holds him in thrall. Again let it be said that the sex act itself, however intimate or acrobatic, need not be obscene to the absorbed pair. Obscenity is an effect of distancing, made possible by a picture or word that turns anyone who uses it into a salacious voyeur, who then redirects that salacious feeling away from himself by saying indignantly, "Look at these groaning, sweaty animals and buffoons!"

Words such as "obscenity" and "pornography" may be objected to on the grounds that they are judgments of the puritanical West, inapplicable to other times and cultures. I wonder whether this is indeed the case. To the extent that any society has a norm of polite behavior, it has a norm of what lies beyond the pale—the risqué, the outrageous, the subhuman, and the obscene. Moreover, I am suspicious of the modern scholar's tendency to regard any sign of disapprobation as Victorian prudery and, at the other extreme, to commend sexual display in art, no matter how extreme, as healthy and frank. To me this attitude suggests an overcompensation for past excesses of moralism that cannot bear critical review. It may also be a device for avoiding frankness where frankness can really hurt. "Be true!" declaimed Nathaniel Hawthorne, not once but three times. "Show freely to the world, if not your worst, yet some trait by which the worst may be inferred."[30] He had, I believe, moral lapses in mind. Total honesty in how we see and treat others is too demanding, but most of us can at least drop nonchalantly a four-letter word or two in conversation by which our animal and sexual nature is acknowledged. The issue even there is how far we can go along that path without dehumanization. Dehumanization does not and cannot mean the return to a prior state of innocent animality. Rather, the process ends in the conjuration of monsters, neither human nor animal, whose reason for existence is to stimulate titillation in private and, more sinisterly, declare to the world that that is what human beings truly are: inconsequential and grotesque. What is a man? To judge by popular Japanese prints of the seventeenth and eighteenth centuries, he is a lascivious beast peeking through refined culture, a penis the size of a bludgeon and the color of a blood sausage exposed under the half-raised silk kimono.

Sex and Fecundity

Since prehistoric times, representations of genitals have symbolized fecundity. Could Chinese ancestral tablets, longtime icon of respectable lineage, be originally phalli?[31] To put it another way, could piety be a refined human sentiment that started its career as fertility cult—the worship of ancestral genital potency? While the Chinese picture remains obscure, there can be no doubt as to the meaning and ubiquity of the lingam in Indian civilization; and, likewise, the visibility, if not ubiquity, of phalli of monstrous size exhibited in public places such as the theater in archaic Greece. Earth goddesses also flourished—outstandingly, Demeter, autonomous queen of the fields, whose ability to produce did not require the attendance of a personified male partner. A significant difference thus emerged in representations of female and male generative power. Whereas the entire female figure stood for potency, in the male it was concentrated in just the scrotum and the penis. This explains why Uranus's amputated male organ could by itself, in union with the sea, create Aphrodite.[32] Men have always been greater patrons of pornographic art and peep shows than women have been. It may be that the cause lies in an important difference in their sexual nature, for compared with a woman's excitation, which tends to be diffused throughout her body, a man's is localized. He is—he is even subordinate to—his genitals, which can seem detachable or to have a powerful will of their own. Fertility cults were once widespread throughout the world. With their decline, the tumescent phallus must sooner or later lose its magical-symbolic role in the generation of life, accepted and even revered by society for that reason, to come to an ignoble end as crude pictures and graffiti on derelict walls, cries for attention and relief by thwarted, lonely individuals.

Fecundity itself has over time taken on an ambivalent odor, except when it is applied to plants and animals. Likewise breeding.

That word, with its jarringly mixed signification of animal urge and a technically guided, manipulative consciousness, is suspect in the human realm. And although many societies, including the highest, very likely had breeding and fecundity (among other considerations) in mind when they matched a young man with a young woman in marriage, shamelessly using all the social forces and skills at their disposal to ensure the couple's conjugation, they could rarely say outright what it was that they were doing; even less could they say outright that the sweaty entwinement of bodies on the marriage bed served merely to effect a material or political advantage. Humiliation could go no further than to be coerced to couple sexually for some other people's purpose. An image of slaves toiling in the field is bad enough, but far worse, not so much in conveying physical stress as in evoking the ultimate in social degradation, is the image of slave farms, such as those established by Muslims in nineteenth-century Dar Fur (Sudan) that specialized in breeding black slaves for sale like cattle and sheep.[33]

"Be fruitful, multiply, and replenish the earth," God said to man ("male and female created he them") (Genesis 1.27–28). Procreation is greatly valued in almost all societies, not least among those that have emerged from the Judeo-Christian tradition. The Catholic church sometimes speaks as though giving birth were the highest or even the only justification for sexual union. But it was not always so. Other values have at various times been given greater prominence. In the Bible itself, God wanted Adam to have a companion. Companionship, characterized as that state for which a man would leave his parents and "cleave to his wife as one flesh" (Genesis 2.24), was emphasized, rather than procreation. Anglicans are taught to believe that "the union of husband and wife in heart, body, and mind is intended by God for their mutual joy." Mutual joy is mentioned first, then mutual help, and, last, "when it is God's will," children. Elsewhere in the solemn ceremony a couple's carnal desire is mutated into a love that is "a seal upon their hearts, a mantle about their shoulders, and a crown upon their heads." Beyond even this image of regal exaltation is the idea that the marriage of man and woman can be a symbol of "the spiritual unity between Christ and his Church" (Ephesians 5.25, 28–30).

Distinctively Human: Eroticism and Love

Sexual insatiability, cruelty, and inordinate violence are uniquely human. So are eroticism and love in all their subtle and passionate forms. They are uniquely human because, for good or ill, imagination is at work, guiding, moderating, or intensifying animal proclivities and impulses. Note how the sex act differs from eating in the measure that imagination comes into play. Eating is pretty much eating, whether humans do it or animals do it, though we humans try to persuade ourselves of a difference by savoring food longer and by putting on a theater of table manners. This is not so in the sex act. Among humans it is never just responding to a physiological urge, but always something much more glorious, but also, unfortunately, much more destructive and sinister. Consider how the tactile sense is activated during sexual congress. Touch matters to humans far more than it does to other animals, including chimpanzees, our closest primate cousins. A well-known reason for the difference is that human beings have strong, flexible, and sensitive hands and that the human body is covered by an expanse of naked, responsive skin, which invites searching caress.[34] A feast of tactile sensations opens up as the lover's hand moves from rumpled hair and firm pectorals to the soft skin below the armpits, the muscled thighs, the knee's hard knob, and as it registers temperatures that range from the nose's dry cool tip to the groin's tropical heat.

Every touch heightens pleasure and desire in oneself and, at the same time, elicits pleasure and desire in the beloved. Seeing this pleasure in the other enhances one's own, and so the pleasure rises to reach a high glow of concordance when the penis penetrates the vagina. Touch shifts at some stage from hand to sexual organ; the uniquely human yields to the sort of coupling that all animals know. Yet it is this act of penetration, rather than the orgasm that floods and drowns all sense of self and individuality, that is the great wonder of human sexual union. At least, according to Rollo May, this is so with lovers who dream about, remember, and savor the experience. To them, entry into the other and reception by the other in desire is a tremulous high point, not erased even by the orgasm that follows, for it is in this first most intimate of contacts that the lovers' sensorial responses are "most original, most individual, most truly their own."[35]

To touch is to be touched, a union in which an individual feels all the more herself as she embraces the world and becomes that world. What happiness to *be* the silken texture of a flower petal, the roughness and weight of stone, the enveloping warmth and yielding ooze of mud! And if this bliss can occur when one sinks into nature, far greater can it be when one sinks into the arms of a beloved. "The warmth of his shoulder against my palm is all the joy there is in life," says a character in a Doris Lessing novel. "I am so happy, so happy. I find myself sitting in my room, watching the sunlight on the floor—a calm and delightful ecstasy, a oneness with everything, so that a flower in a vase is oneself, and the slow stretch of a muscle is the confident energy that drives the universe."[36]

What of the human face? I have avoided mentioning it thus far because I wanted to start with the animal self, skin rubbing against skin. So, what of the human face? It is by far the most expressive part of the body, an expressiveness recognized and enhanced by the world's art and literature.[37] The face is the index of the whole person—the heaviness of earth, the empyrean reach of mind, and the profundity of soul. A face commands love. "Love" seems the right word to use, whereas with other parts of the body we are less certain. Can a hand command love? Perhaps, for the human hand can be very individual and expressive. An arm, a shoulder, a foot? "Fetishism" or "lust" comes to mind when interest in them is irresistible. How natural it is to have the face of the beloved on the mantelpiece, and weird for any other part of the anatomy to be similarly captured and displayed! And it goes without saying, even among the most libertine, that a snapshot of someone's sexual organ taken for purposes of private viewing and arousal is obscene.

The face, perhaps because it suggests the whole person, projects power in a way that the rest of the anatomy does not. From one perspective, it is most vulnerable—the eyes in particular, which as "windows to the soul" imply total exposure. And yet from another perspective, the face—and again the eyes in particular—commands a world. To see a face is to see a subject rather than an object. Think how untrue this is of any other part of the body and of the body as a whole when it is asleep or unconscious. The sex organ also projects power and vulnerability, but in a totally different way. Power in the sexual organ is raw and unmediated. Curiously, that is also the source of its vulnerability. The mere presence of another can cause the penis to stiffen uncontrollably. Whereas the face can deceive—feign and simulate—the penis cannot, and therein lies its innocence.[38]

Humans fall in love. "Fall" suggests the involuntary. Falling in love "at first sight" underlines the involuntariness even more. But when one considers the complexity of the emotion, this manner of speaking is hyperbolic, justified, if at all, by the force of the emotion and its element of surprise. Falling in love, as Irving Singer has noted, is more conceptual than instinctive, more a disposition prepared by a mind that has come under the influence of society and art than an innate bias and surge of hormonal pressure.[39] Of course, the disposition has to be activated; otherwise, it remains a diffuse yearning. When it is activated—triggered perhaps by a single encounter—the searing emotion seems to come out of the blue, a moment in time that marks the beginning of a new life. The past becomes "prehistory."[40] Lovers, liberated from the rancors and resentments of their earlier lives, find it surprisingly easy to forgive the people who caused them. A past that no longer taunts is a sunny landscape that invites the lovers, when they are not wholly engaged in the present or planning for the future, to stroll through hand in hand, pausing here and there to honor a fond memory.

In love, the beloved is seen to possess not only beauty and goodness but even delectable weaknesses and faults that are found in no other individual. "Let the king have sixty queens, eighty concubines, young women without number! But I love only one," sings Solomon in the Song of Songs (6.8–9). Romantic passion is not unique to the modern West, as was once thought. People have fallen in love and idealized the beloved in all times and places—in China and India, Morocco and Kenya.[41] But it may be that one state of love is more strongly developed in the modern West than elsewhere. It is a state that makes extra demands on the imagination, for the individuality it recognizes in the beloved is not just a virtue like courage or a quality like blue eyes; rather, it is the loved one's unique way of perceiving and living in her world. A man may well think that he "owns" his beloved's blue eyes; he cannot so easily think that he owns the beloved's experiences—her reality. He may enter and enjoy it, but only as a privileged guest.[42]

Sacred and Profane Ecstasy

In love, one feels simultaneously vulnerable and strong, united with the beloved and isolated in a heightened sense of self, abased and exalted—delicious emotions that well out of one's innermost being and gratitude for the splendors of the objectively real. Almost the same words serve to describe an encounter with the divine. To most people, the tidal swing of emotion culminating in the ecstasy of sexual embrace is the most they will ever know of mystical union and transcendence. Not surprisingly, world literature and art—-notably those of India, China, and the West—have seen fit to conjoin the two. There was then no cover-up, no sense of a split between body and soul, the "low" and the "high." In the absence of such a sense of a split, a body evokes feelings that merge and rise effortlessly to the realm of the spirit. In the West, Solomon's poetry (Song of Songs) provides an early and familiar model of how a charged erotic language can call forth the passion and mystery of divine union. Again and again, Western mystics and poets have availed themselves of this model, outstandingly St. John of the Cross, whose Spiritual Canticle contains images that could shock even W. H. Auden, a very worldly and modern poet.[43] In sculptural art, the voluptuous forms of gods and goddesses, the transparent identification of sexual congress—the herd girls' desire for Krishna—with mystical ecstasy, richly adorn the exterior of Hindu temples.[44] Almost as gloriously unselfconscious are works of Western sculpture and painting from classical antiquity to the seventeenth century. Over and over again, the perfectly formed human nude is made to serve as a symbol of spiritual perfection. In medieval times, sculptors saw nothing amiss in using pagan Nereids to represent blessed souls on their way to heaven. During periods of religious fervor, outstandingly the sixteenth century, the line between sacred and profane ecstasy was exceedingly fine. Saints, daringly uncovered, turned their eyes to heaven in bliss. Michelangelo's drawing of the risen Christ (housed at Windsor) shows him completely nude, with exposed genitals. To the art historian Kenneth Clark it is "perhaps the most beautiful nude in ecstasy in the whole of art."[45]

Transcendence

Is it really possible to look at the image of a beautiful human body and not be sensually (sexually) aroused? Gerard Manley Hopkins found the naked Christ a strong temptation to erotic daydreaming.[46] One may dismiss Hopkins as a repressed priest living in Victorian times, with an imagination that could be inflamed by the most innocent matter, but I ask again, Is it really possible for the dense, sexually charged human body to be a window to spiritual values, as words on the printed page are such a window to ideas? Symbols of a spiritual realm are least likely to divert the imagination from their intended path when they are themselves abstract (as, for example, the circle and the cross, light and darkness, the vertical and the horizontal) or when they are taken from nonhuman nature (mountains and forests, the lion and the doe). But the naked human body, unless it be that of an infant, is none of the above. It is in a category of its own, even though it belongs to nature.

A strong tradition exists in the West, and in other civilizations as well, that is deeply suspicious of the body, which is considered—depending on the tradition—the fount of intemperance and lust, of ignorant materiality and illusion, of the sickness, suffering, putrescence, and death inherent to biological life. As we have noted, the ancient Greeks entertained no such bias. To them, a perfectly shaped human being (outstandingly, a naked boy) could be an image of the Good, the only aspect of the Good (the aesthetic aspect) to which humans were naturally drawn. But a puritanical strain also existed in Greek thought. It was given prominence by Plato, and it warned that one must not dwell on the body's aesthetic-erotic appeal, for that way lay bondage and corruption; rather, one must turn one's attention away from a particular person's fairness to fairness in all persons, from fair forms to fair practices, from fair practices to fair ideas, and ultimately to the idea of absolute beauty and the Good. These are the famous rungs of Diotima's ladder. Progress is upward, from the particular and material to the general and abstract, from decay and change to unchanging perfection, from image to reality (*Symposium* 211–12). The ladder is a means of escape. Each step up takes one closer to the real.

The human body received respect from the Greeks and even from an important line of Christian thought, which viewed it as a "temple of the spirit"—a suitable vehicle for divine incarnation. Nevertheless, another line came to dominate Christendom, one that looked upon the body, its concupiscence, with deep ambivalence if not outright hostility. Paul has often been blamed for this second line, and he earned the blame not because he thought sexual union within marriage wrong but because he tended to consider the institution of little importance when the Kingdom of God was so close at hand. Devotion to spouse and offspring, commendable in its own circumscribed sphere, distracted one from the higher calling. Continence was the answer. Continence was a gift that made it possible for Paul and others similarly endowed to attend to God's business wholeheartedly. Paul knew that they were a small minority, that most of their converts could not abstain from sex, and that for them it was "better to marry than to burn" (1 Corinthians 7.9). Clearly, Paul deemed married life second best. However, as his frustration over human frailty increased, he went further. He took to substituting the derogatory word "flesh" for "body" and then spoke of "flesh" as the opposite of "spirit."[47] Paul's shift of attitude had major consequences for Christian moral teaching. If, for Paul, the shift contained an element of inadvertence, there was no such inadvertence in the carefully weighed judgments of the young church's three towering leaders, who in prestige and influence came immediately after Paul: Ambrose, Augustine, and Jerome. Together they managed to bind sex firmly with sin, sin with corruption and death, and, conversely, virginity with heroic virtue and eternal life.[48]

Dying and Death

As animals, we die—a plain factual statement that is nevertheless riddled with paradox, for what animal will call itself an animal and then go on to say that as such it dies? Can an animal that knows itself to be one die? Doesn't this "knowing" hint at a level of being that just might survive bodily decay? I have noted that eating and sex are an embarrassment and a moral dilemma because we are able to stand outside ourselves, watch, and ponder. Is the same true of dying and death? The answer is not so clear. The sort of reflection that makes eating and sex a problem also makes dying and death a problem. But there are surprising differences. Consider the matter of being watched. One does not like to be watched while one eats and even less while one engages in sex. But dying? If I thought I were dying, reduced to the dependent state of a child, wouldn't I welcome watchful eyes? My human worth, I should feel, would be enhanced rather than diminished. In sharp contrast to sex watch, which is perverse and degrading, death watch—vigil beside a dying or dead person—is a good custom, an achievement of culture; indeed, turning one's eyes away from the dying and the dead is reprehensible.

As experience, eating and sex differ radically from dying and death. Dying, unlike eating and sex, is normally gradual and nonrepetitive. Dying may be so gradual that we are not even aware of it happening—we do not know it as experience—until the critical stages toward the end, when pain first discolors and then consumes our being. As for death, it is not itself a personal experience—not an event that one can ever "go through." It is "the one experience I shall never describe," said Virginia Woolf, who so expertly described the death of a moth.[49] Nevertheless, human imagination can dwell on one's own death such that a mere endpoint is given the power to haunt or color the whole of life.

Death's Shadows

Consider some of the ways that death haunts life. I use the word "haunt," as distinct from "color," to suggest the negative, and it is with the negative that I will start. It can be assumed that the intensity and extent of the haunting depend on such factors as an individual's temperament and propensity to reflect, and the culture in which he or she is raised. Culture (group value) may indeed be the paramount factor. However, rather than take up variations in degree of haunting among individuals and groups, I should like to lay the philosophical groundwork by turning to certain common human experiences and tendencies.

Death, I have noted, is not something one can know directly. "Where I am, death is not; where death is, I am not," as Epicurus put it.[50] On the other hand, we easily see it happening to others. A person alive one day may be dead the next. What happens to another can also happen to me and to those I hold in affection or esteem. The glow of life is admirable, yet it cannot last. Moreover, is it even entirely admirable? I value my own life and its sensory satisfactions, but these depend on the deaths of other forms of life. If I do not do the killing myself, I know it occurs next door, and I participate in their ultimate reduction by eating them. This is a commonplace of observation and experience, not an esoteric truth. Food that supports life and conjures so many happy images is itself a richly flavored compost of death. I touched on the moral dilemma of food earlier and will not repeat myself here. But what about sex, that second compelling demand of our animal nature? It can give life and is not obviously shadowed by death. Yet life and death are inseparable. Their odors are mixed or waft uncomfortably close. In human birth it is hard to decouple, in fact and in the mind, the emergence of life from the expulsion of waste. The sex act itself—ejaculation that entails a sense of loss, the phenomenon of postcoital sadness—hints at individual mortality even as one generation is engaged in reproducing another.

In growth there is much passivity—significant loss even as more is gained.[51] In the rising arc of life, one may be aware only of the gain—of increasing competence and a widening world such that the passing of earlier selves and experiences, powerful and wondrous in their own way, is scarcely missed. In the cycle's downward path, the losses mount and a time comes when they can no longer be ignored. The world contracts, or, to put it another way, people, things, and places withdraw—move away, leaving one isolated, abandoned. Death is imagined as this final isolation and abandonment, but its shadow falls on life long before the event itself arrives. The haunting may be mild or severe, depending on individual temperament and the cultural mood of the time. For an example of severe haunting, I turn to China; and I do so because it seems to me that Chinese society, lacking the comfort of either a triumphalist religion or an austerely stoical philosophy, tends to fall into bathos when death occurs. The poet T'ao Yuanming (365–427) captured the bathos in a poem composed in the style of a coffin-puller's song *(wan-ko)*, so called because it derives from the dirges sung by the men as they pulled the hearse to the graveyard. However, rather than put himself in the position of the mourners, which would have been depressing enough—though redeemed by the prospect of the funeral feast to come—the poet took the viewpoint of the one who had been left behind:

> Those who just now saw me off
> have all gone back, each to his home,
> my kin perhaps with a lingering grief,
> but the others are finished with their funeral songs.
> And what of the one who has departed in death?
> body left to merge with the round of the hill?[52]

Losses natural to growth are not usually regretted. All over the world, people celebrate the passing of childhood and the attainment of maturity. True, pain, sickness, and permanent debility can occur at any stage of life. Even in the bloom of youth one may experience severe diminishment, a foretaste of death. But these facts do not instill terror. Life is potently regenerative—ever hopeful, and for good reason: One does recover and can no more feel the excruciating aches and pains of the past that one can recapture the sensation of icy cold in the midst of summer. Suffering is thus a misleading memento mori. It exhibits duration and it passes, totally unlike the endpoint that is death. Moreover, suffering, because it is, after all, life and even a rather intense form of it, does not prevent—indeed, it actively encourages—daydreaming. Watching flickering images in the mind is a ready means of escape, and why not, if it relieves temporary distress? Daydreaming can, however, become addictive. One learns to be not just an occasional visitor but the habitué of a fantasy world. Of that world's many delusory assurances, perhaps the most cunning is the excision of finality from death, making it seem like mere suffering.

So, rather than sickness and suffering, accident (Iris Murdoch suggests) is a better reminder of death.[53] The total passivity that death imposes has its closest analogue in accident, which by definition is something that happens unexpectedly. Accident is statistically inescapable. When it occurs, though life may turn for the better, more often it seems to turn for the worse—to precipitous decline that may end in death. Tribal societies are known to confound the two; death is itself an accident rather than a biological inevitability.[54] Death ought not to occur, yet it does—perhaps suddenly, when one least expects it. Within minutes of eating tainted food, one becomes a sick dog or, more tragicomically, a corpse ready to be swept out of the restaurant along with the other waste. Accident is the most authentic foretaste of death's imperium—fate that is beyond imagination's power to cover up. Most accidents are, of course, minor. But even those—even a stumble—can jolt us rudely from our assumption of life's even flow.

Death's Gifts

Death, surprisingly, also has gifts. It consoles, it gives an extra edge to life, and it is the ground of virtue. Death consoles if one thinks of it as the portal to heaven—a familiar route of escape to many in the Christian and Islamic faiths. Death also consoles *because* it is the endpoint. Jorge Luis Borges said of his aged and sick mother that she would wake up in the morning and cry because, contrary to her fond wish, she was still alive; escape had once again eluded her.[55] Asked to account for his longevity, an octogenarian replied, "Oh, just bad luck!" David, the boy who lacked immunological defense and had to live in a sterilized bubble his entire life, said a short while before his death at the age of twelve, "Here we have all of these tubes and all of these tests and nothing is working and I'm getting tired. Why don't we just pull out all these tubes and let me go home?"[56] Among death's uses, the philosopher Sidney Hook said, is that

> it gives us some assurance that no evil or suffering lasts for ever. To anyone aware of the multitude of infamies and injustice which men have endured, of the broken bodies and tortured minds of the victims of these cruelties, of the multiple dimensions of pain in which millions live on mattress graves or with minds shrouded in darkness, death must sometimes appear as a beneficent release not an inconsolable affliction. It washes the earth clean of what cannot be cleansed in any other way. Not all the bright promises of a future free of these stains of horror can redeem by one iota the lot of those who will not live to see the dawn of the new day.[57]

Death enhances life by confronting us with an Absolute. Without it, not only can this life seem to be just "one damn thing after another," but the afterlife as well—presumably "one damn good thing after another" to all infinity. "I don't look forward to an eternity of survival," wrote the distinguished philosopher Karl Popper. "On the contrary, the idea of going on for ever seems to me utterly frightening. Anybody who has sufficient imagination to deal with the idea of infinity would, I think, agree." Death, he believed, "gives value, and in a sense almost infinite value, to our lives, and makes more urgent and attractive the task of using our lives to achieve something for others, and to be co-workers in World 3 [that is, the world of knowledge and art], which apparently embodies more or less what is called the meaning of life."[58] John Cowper Powys, that maverick Welsh English writer, had the same essential idea, although he put it somewhat differently, with greater emphasis on happiness as the incontrovertible, great human good. To Powys the line in Wordsworth "The pleasure which there is in life itself" rings true, provided it is expanded to include a clear reminder of death. Happiness? "I tell you the foundation stone of all human happiness is the thought of death."[59]

As for virtue, what is it without the risk and terror of death? In the words of Oliver St. John Gogarty,

> But for your Terror
> Where would be Valour?
> What is Love for
> But to stand in your way?[60]

Valor and love are two positive virtues, but a keen awareness of one's own finitude also produces at least one admirable negative virtue, namely, a certain indifference toward the pomp and circumstance of this world. Malcolm Muggeridge in his old age wrote, "Now the prospect of death overshadows all others. I am like a man on a sea voyage nearing his destination. When I embarked I worried about having a cabin with a porthole, whether I should be asked to sit at the captain's table, who were the more attractive and important passengers. All such considerations become pointless when I shall soon be disembarking."[61] In the voyage of life itself, death's shadow is ever present, not just something that haunts the end; at any moment I may receive notice to disembark. Grasping this fact should persuade me that there are better ways to spend my allotted time on the boat—watching the night sky, getting to know fellow passengers, reading good books—than worrying over material advantage and prestige.

Beyond the Endpoint: Nothingness, Gloom, Corruption

Mystics may say that at death we fall into the great Abyss, Darkness, or No, which is God. But this is unprovable, and tagging on God at the end sounds like a comforting verbal trick. The part of the assertion that we *can* be certain of, says Karl Barth, "is that man is negatived, negated."[62] Negated—annihilated. Is nonexistence itself the ultimate dread that humans cover up with such ingenuity? The following grim conversation took place on April 15, 1778, between a Miss Seward and Dr. Samuel Johnson, when Johnson was sixty-nine years old: "Miss Seward: There is one mode of the fear of death which is certainly absurd; and that is the dread of annihilation, which is only a pleasing sleep without a dream. Johnson: It is neither pleasing, nor sleep; it is nothing. Now mere existence is so much better than nothing, that one would rather exist even in pain than not exist."[63] The void awaits everyone. As I move toward that inevitability, it can render all my life, its struggles and rewards, shadowy and ultimately meaningless—a long preparation or education (as William Butler Yeats put it) that ends nowhere. John Wesley, it seems to me, had something like this collapse of meaning in mind when he confessed, in a letter (1766) to his brother Charles, "If I have any fear, it is not of falling into hell, but of falling into nothing."[64] Hell at least means that what I do in this life has lasting consequence; every moment may be momentous. But what if there is neither heaven nor hell, not even some sort of perch in the afterlife from which I can see in joy or sorrow the effects of my own earthly existence? Does then everything—not only seating at the captain's table but also savoring the night sky—become a more or less successful distraction before the great No?

The human mind cannot come to grips with "nothing." Therefore, even when people imagine the afterlife negatively, they do not see it as nothing but as something—a shadow of this life. The drearier they imagine the underworld to be, the more they appreciate sunny well-being in the world above ground. A case in point were the Sumerians. What did they treasure? "Wealth and possessions, rich harvests, well-stocked granaries, folds and stalls filled with cattle . . . successful hunting on the plains and good fishing in the sea."[65] In short, substance to counter the lack thereof beyond. Ancient Hebrews too conceived of Sheol as a shadowland, which was one reason why they so warmly embraced this life's blessings. People in general want to enjoy material abundance. "Good," to nonphilosophers, is something tangible that benefits life; in most societies it also connotes "abundance"—initially the plenitude of nature, then the plenitude of manufactures. Notable exceptions occur in certain specialized cultures—hero culture, for one. Warriors in ancient Greece certainly appreciated good things, which they might fight for, but they valued honor more, which they sought in the teeth of death. Yet their idea of the nether world was so dreary that, to escape it, even honor might be forgotten. How the horror of bare existence could humiliate even the bravest! In a picture made famous by Homer the dead are likened to bats fluttering and screeching in a cavern, and the ghost of Achilles is heard to say that he would rather be the serf of a poor man on earth than rule over all the dead.[66] Even the serf can bask in sunshine, quench his or her thirst with spring water, bite into a loaf of bread. And what if one has more than that—a good job, family and children, friends—as many people do? The loss then is immense. We can understand why the novelist Joyce Cary, as his health steadily deteriorated, broke down. "His grandson, Lucius, at Joyce's request, sang 'O Little Town of Bethlehem,' standing in the open doorway of the bedroom so as to be able to hear the piano. When it was over, Joyce burst into tears. The injustice of life and the knowledge of what he was losing had overcome him."[67]

Death is quickly followed by the indignity of corruption. But rotting can occur long before death. Freud's cancer of the larynx produced an evil odor that offended even his faithful dog. He and others around him could smell the approach of the endpoint before it arrived. The entire civilizing process may be seen as an effort to bury the fact of death and its premonitory signs—removing from genteel eyes and sensitive noses the necessary work of the butcher, the disposal of animal and human corpses, the care of the sick (not always a sightly undertaking), and all odors of decay, which were once believed to be a potent cause of death, whether as miasma from swamps packed with rotting organic matter, or from city burial grounds swollen with human corpses, or from the filthy, densely packed quarters of the poor. "Cleanliness is next to Godliness." That ancient saying revived by John Wesley may have a meaning other than the one normally given it: a clean, odorless body is a mark of immortality.

Immortality

Contrary to common belief, it is not the case that people in earlier times generally expected to survive death in the form of resurrected body or spirit to enjoy the rewards of paradise. Ethnographic and historical evidence suggests that the vast majority of human beings were too humble—too beaten by life's recurrent insults and injuries, and too habituated to living in dire need surrounded by filth—to postulate living in heaven in bodily splendor.[68] For personal survival in style to be conceivable, there must first be a strong sense of individuality, of being a person who in this life can already engage in enterprises of worth and valor. Among quasi-egalitarian communities, great hunters such as the Iglulik are well known for their belief in a happy afterlife in the Land of the Moon.[69] This is understandable, for hunters are individualists; in their search for game they must often take initiative alone, or with a partner, and they must often confront danger bravely. By contrast, cultivators and peasant farmers are more group-oriented, the demands of their livelihood providing few opportunities for individual initiative;[70] and perhaps for this reason personal immortality, as recompense for meritorious deeds done here below, is not an element of their world-view. In hierarchical societies, at first only the topmost crust of rulers and warriors had enough presumption to postulate a glorious afterlife. Later, as the livelihood of people in the lower ranks improved and there was more opportunity for them to take pride in their occupations and in themselves, they too found it psychologically compelling to assume the continuation of a richly satisfying existence in another world.

What is bliss like in the thereafter? The human imagination has repeatedly failed to come up with a worthy image. One view, already noted, is that the life beyond is a shadowy existence and not bliss at all. Even when that world is given greater substance, as in many lineage societies, it is still a pale replica of this one. In such a world, ancestors are provided with food and other necessities, are addressed as though they were still alive, and are housed in familiar buildings, but how they live and exactly what they enjoy are seldom described.[71] Vividly original images of the afterlife are far more likely to pertain to hell than to heaven. The human mind seems better at conjuring horror than bliss. The one has the punch and texture of reality; the other is merely a place we already know purged of pain and sordidness, a pastoral landscape or an Islamic garden of delight, in which "the blessed shall recline on soft couches, be served with a goblet filled at a gushing fountain . . . and sit with bashful, dark-eyed virgins, as chaste as the sheltered eggs of ostriches" (the Koran). Theologians must work hard to give a figurative twist to these sensual/erotic images. At their most convincing, evocations of heaven avoid material voluptuousness and resort instead to the abstractly aesthetic and intellectual, as in Dante's paradise, or to dances of light, color, and sound, nonrepresentational yet of immediate allure to the senses, as in the interior of a great cathedral.[72]

Escape is a response to both push and pull. In deciding to cross the ocean to live in the New World, migrants may well have been more pulled by its paradisiacal reputation than pushed by the miseries of home. The paradise promised beyond death, however, can never have been a serious pull. People of sound mind do not commit suicide because they have conjured an afterlife of irresistible allure. Strange to say, the idea of the void can be a real draw, and this even to individuals who are not *in extremis* but who have a deep longing for things (even good things) to come to an end at a ripe time. The absoluteness of the end has an aesthetic/moral appeal of completion and integrity that the pie-in-the-sky models of heaven lack.

Stoics in classical antiquity embraced an austere outlook on life that included a dignified way to bring it to a close. Well known for their love of fate (with its implied passivity), they nevertheless believed in control over the self. Virtuous living here and now was to them the proper aim, for it alone confers dignity and honor. By contrast, attempts at envisaging the beyond are exercises in debilitating fantasy. People of a naturally stoical temperament or who have disciplined themselves to become stoical have no doubt existed in other ages, including our own. But it is safe to say that they have always been a small minority, and especially so in our time. Why? Let me offer two reasons. First, postulating some sort of survival, if only as an anxious eye watching to see how our biological or intellectual progeny are doing, has by no means weakened even among committed modern secularists; perhaps it is an irreducible element in human mental makeup that cannot be entirely erased whatever one's professed belief. More to the point, for it highlights an important difference from the past, is modern men and women's undoubted antipathy to the Stoic idea of fate, especially when this "passivity" is paradoxically combined with the demand to exercise heroic control over the self—the body and its passions, the mind and its wayward imaginings. Fate to modern men and women is an outmoded idea even though they continue to experience it under the name of "accident." As for control, they fully endorse it, but to them, in contrast to Stoics, control means keeping the body perpetually fit rather than keeping desire and will temperate and wise. The body attracts more and more attention now because for the first time there is reason to believe that medical science can maintain it far beyond the "normal" span.

How long can a human life last? Is there a theoretical limit? Can science postpone death forever? As yet few people take the last question seriously. It still sounds more like science fiction than science. The question also deeply offends certain moral-religious scruples. Christianity's hold on the West may have weakened greatly, but it hasn't disappeared; the idea of maintaining the human body indefinitely is still too radical a departure from the long-held view of its return to dust and ultimate resurrection. Moreover, theological details apart, there is the lingering awareness that human beings are creatures. Their power is or ought to be limited; and so the idea that they can one day prolong life forever seems the ultimate usurpation of divine prerogative—a transgression that can have the direst consequence.

Conquest of Nature—and of Death

In the twentieth century, a critical and even alarmist attitude toward technological accomplishment is common in intellectual circles of Western Europe and North America, parts of the world where the greatest technical advances have been made and life is generally the most secure and affluent. Indeed, in recent decades the Atlantic West has moved so far from the hubris of an earlier time that to its more fervent environmental advocates even the expression "conquest of nature" is avoided. Nature has come to mean almost solely organic life—a thin and frail mantle that has been abused and that now needs human protection. So, if we want a glaring example of Western hubris in the twentieth century—Western in the larger civilizational sense—we must look eastward to the former Soviet Union, where "conquest of nature" remained the official doctrine until the 1980s.

Nature is life and death, nurturing and death-dealing. Historically, it is the niggardly and death-dealing aspects of nature that have been uppermost in human consciousness. Whether as an oppressive background presence or as an immediate threat, these aspects cannot be ignored. Much of the human story is one of trying to escape the oppressiveness and the threat. Deflecting the ferocities of weather, ensuring water for crops, promoting livestock and human fertility, maintaining an adequate supply of food, were and are the basic businesses of society. Repeated success in these areas constitutes progress. Nature's constraints can be removed by small steps—or, thanks to science, by leaps. An optimistic view sees vast tracts of sterile land turned into unending fields of corn through feats of river diversion and irrigation; and, more recently, it sees genetic technology gaining direct control over the forces of life. Such a rosy prospect, challenged in the Atlantic West, is fully conformable to Leninist-Marxist ideology and so continued to be upheld by the Soviet government until its collapse.

In the midst of Soviet optimism an awkward question intrudes, namely, Just how far can this conquest go? For when we probe into what all the human struggle is about, it is ultimately the overcoming of death. Can humans become immortal? Guardians of Leninist-Marxist ideology deflect the question by saying, Yes, they become immortal in their works and achievements and in the memory of future generations. But this is mere common sense. The hard question is individual immortality and, should death occur, the possibility of recovery—or, to use an older word, resurrection. Defenders of official thought, although they find the idea of personal immortality distasteful, cannot deny it outright, for to do so would be to put a theoretical cap on the conquest of nature. There exists, then, a loophole in the communist Faustian ideal. Through it, certain Russian writers and poets have introduced a Christian language of hope.[73] In the communist utopia there will be no slums or poverty—and presumably no filth, disease, or avoidable pain. If the list of evils that can be overcome stops here, one may still say, Yes, this project is hugely ambitious but not impossible. However, to stop at any point is to admit human impotence, which the Leninist-Marxist doctrine doesn't permit. So the list must go on. Does the future also exclude decay, the sewer odors of mortality? Death? The poets' answer, more wild-eyed than even that of party ideologues, is yes. Their utopia is not only tractors and plentiful harvests, happy children romping in the park, social justice. It is far more; it is no less than the Garden of Eden and the Celestial City combined, a world in which Lazarus can walk out of the tomb and Christ can emerge from his death shroud. The critical difference between these religious images and those of the secular, utopian future is this: Whereas in the past the images were mere images and poetic flights, in the future they are detailed plans for the construction of handsome buildings and gardens; and, more generally, they are rational procedures for overcoming poverty and disease, and, beyond these specific ills, nature at its most sinister—decay and death. Can the dream of escape from our biological condition—our animality—go further?

3

PEOPLE

Disconnectedness and Indifference

• • •

I have drawn attention to culture as escape from nature, and escape into "nature" itself as escape into culture—that is, a world of culturally derived meanings encapsulated in such words as "countryside," "landscape," and "wilderness." I have also drawn attention to nature not as an external environment but as our own body, and told the story of how as cultural beings we seek to escape that animal self, which, though a self, can also seem to be the Other.[1] I now turn to a third aspect of our existence, which is rooted in our biological nature but not confined to it. This is the uniqueness of each and every human individual, a uniqueness that makes for a sense of isolation or disconnectedness even in the midst of familiar others; and that disconnectedness is, albeit rarely, ground for feeling the world's profound indifference—its aloof and essential Otherness.

By interpreting culture as escape, I have perforce given it a dynamic meaning. Attention is directed to how people make plans, tell stories, perform ceremonies, migrate, transform. Culture is thus activity. But it is also the end result, or the product, of that activity. The product—story, ceremony, house—conserves, stabilizes; it becomes in time an environment and a routine that can be taken for granted. Culture in this sense enables us to forget that we ever sought to cover up or to escape from the Other; it creates an atmosphere of ease; it makes us feel that "we are fine as we are, where we are." Ease, so fundamental to our sense of well-being, presupposes a shortness of memory—the ability to forget. Culture enables us to forget the menacing Other—weather, for instance—by constructing a house. But its most basic contribution to forgetfulness, to an individual's sense of ease, lies in the construction of a "we." "I" may be feeble, but "we" are strong. Historically and all over the world, "we" is the preferred pronoun. "I," by contrast, is seldom used. Only from the sixteenth century onward, in one part of the world (Europe), did "I" gain a certain cachet. In modern times a substantial number of people, conscious of their ability to order their own life and even rearrange reality out there to their taste, use "I" proudly. Still, the moral odor of "I" is suspect. It carries undertones of selfishness and aggression. Contrariwise, "we," as in *we* Americans, *we* environmentalists, or *we* the people, is confidently righteous.[2]

Uniqueness

An isolated individual is vulnerable, and we can understand why it seeks the strength of the group. A human individual, however, is also highly distinctive, even unique. Is not uniqueness a quality that all humans take pride in? The answer is, Yes and no. Uniqueness is problematic everywhere. It is a state of being that a person both wants and does not want, although the degree of desire either way varies with culture and individual temperament. In the United States no one wants to be treated as part of the woodwork. Most Americans like to be recognized for their distinctive quality, their importance to the group; they also like to be rewarded accordingly. In other societies, especially those called folk or traditional, people are less eager to stand above the crowd. An individual in such a society would not want to be seen as assertive, which could make him a target of criticism (if his initiative led to unrest) or of envy (if it led to exceptional achievement). Yet in more covert ways each person still wants to be acknowledged as somehow special and nonexchangeable with others. This wanting to be special and so deserving of notice—a close cousin to respect—is one face of human nature. The other face, as rooted in human nature, is the desire, even among ardent individualists, to fade at times into the background. To stand out is ego boosting, but it is also tiring and stressful. It not only exposes one to criticism, it can also lead, as a consequence of isolation from the shared practices and values of the group, to attacks of melancholia and a sense of meaninglessness.[3]

Individual human uniqueness has a variety of causes. Among them, at the most basic level, is biology. In a cosmopolitan city, anyone can see how humans differ in body size, shape, and color, often to an arresting degree. More subtly, body scents and fingerprints vary from person to person. Outwardly so unlike one another, humans are hardly "brothers under the skin." On the contrary, as the biochemist Roger Williams noted, the sizes and shapes of stomachs differ far more than do those of noses and mouths. If noses varied proportionately to stomachs, some would look like cucumbers, others like pumpkins. A hand with six fingers is considered abnormal, yet the pipes that branch from the aorta above the heart vary in number from one to six. Individuals equipped with a narrow esophagus have a difficult time swallowing pills; at the opposite extreme, those well favored may accidentally ingest a whole set of false teeth. Ambitious politicians would do well to have an esophagus large enough to allow them to finish eating and still leave plenty of time to bend their neighbors' ears.

The perceptual senses vary widely in their scope and degree of sensitivity, even among individuals considered normal. Ears that can barely register sound at certain frequencies may be supersensitive at others. When pitch sensitivity is combined with other powers of hearing, people's daily lives are affected in matters such as toleration for the level and kind of noise, the ability to hear certain words but not others, and the appreciation of music. Standard eye tests reveal unique strengths and weaknesses in visual acuity. Peripheral vision is not ordinarily tested, yet it can show differences that affect competence in sports, driving cars, or flying airplanes and perhaps also in the ease and speed of reading.[4] Whereas color vision is a species trait, sensitivity to shades of color and to the appreciation of a color's richness can differ widely. "Is the redness of the rose the same to you and me?" lovers may wonder as they stroll hand in hand through a garden. But the question is not merely philosophical; it is also neurological—a matter of knowing the number of pigment genes on the eye's X chromosomes.[5]

Most remarkable of all are differences in the human brain. Every feature that has been measured shows unexpected diversity. The brain makes every individual truly unique, a fact that is a source of pride but also of isolation and discomfort. A wizard at chess may not be much good at algebra. Excellence in one branch of mathematics does not guarantee high performance in another. The talented French mathematician Jacques Hadamard admitted that he had difficulty mastering Lie group; it was as though his mental energy for that specialty had been exhausted in the process of attaining mere competence.[6] Some people are very verbal, but there too the talent may show itself in one area rather than another—for example, in poetry rather than in expository prose. Double negatives in a sentence can be a stumbling block for some listeners who otherwise do not lack aural-verbal competence. Exactly how specialized is the human ability to form grammatical sentences? An extreme example is a family whose members, as a result of a defective gene, stumble over plurals, although in other respects they speak or write with normal fluency.[7]

People's Indifference

One consequence of human biological uniqueness is that a person often feels slightly out of step with other persons, including ones who are closest by virtue of blood or affection—still eating when others have finished, feeling cold when others complain of heat, unable to catch the meaning of a sentence when others nod, and so on in the course of an ordinary day. Reminders of disconnectedness are, for obvious reasons, repressed whenever they appear, by both the individual and society. Lessons start early. In the family, members are schooled to attend to what holds them together. Although each member lives in a world uniquely its own by virtue of age, sex, ability, and temperament, the reiterated message is that family members are one, knitted together by common purpose and complementarity. Still, it is within the family that a child first learns about human disconnectedness. The sense of disconnectedness may be a consequence of the sort of biological differences I have just noted. But this is not necessarily the case. Indeed, the contrast in size between child and adult, rather than opening a chasm, sets the stage for the child to climb onto his mother's lap and fit himself snugly there to mutual satisfaction. No, what brings disconnectedness to the forefront of consciousness is far more likely to be a conflict of intention and project. The child learns that, just as he has a project suited to his age, talent, and experience, so does his mother have one suited to her age, talent, and experience, and that she is not always willing to abandon her project for his. The look of annoyance that clouds the mother's face as the child disrupts her reverie to show her a picture he has proudly drawn creates in him, however fleetingly, a feeling of shock and disorientation.[8]

Such a dispiriting commonplace is usually brought to our attention by someone sensitive and articulate, whom society may not especially welcome, for from its point of view many things are better left hidden. When, as members of an enlightened society, we expose the painful qualities of such things, we should do so with circumspection, remembering a human being's psychological vulnerability, which is rooted in the young human's prolonged dependence on other people. The nature of this dependence changes with growing maturity, but dependence of some kind remains a problematic and often humbling relationship throughout life. To the one example I have already given, let me add a few more so as to suggest the scope and range of situations in which the failure to find the expected response—nothing dramatic, yet additively oppressive—occurs in day-to-day living: An infant struggles to find the nipple of a mother who has drifted off to sleep. A room suddenly elongates when a child realizes that the object of his infatuation has merely played with him.[9] An office worker has the sense of being out of step with her group, no matter how hard she tries to belong. A look of absentmindedness in the listener reveals the storyteller's own enthusiasm as naive. Tiny acts of betrayal and fleeting sensations of being betrayed darken the daily marketplace of love.[10] One feels that human relations are ultimately what matters, yet this feeling is combined with the chilling knowledge that, as Albert Camus puts it, "it is only our will that keeps these people attached to us (not that they wish us ill but simply because they don't care) and that the others are *always* able to be interested in something else."[11] Last, and as a sort of shorthand, there is general truth in the cliché "out of sight, out of mind." We strive to remain "in sight" while knowing deep down that death can remove us at any time—once and for all—from sight and, soon, from mind.

Plants and Animals

The world is composed of people, other living things, and inanimate matter. If people at times find their own species, their own kinfolk, uncomprehending and incomprehensible, what expectation can they reasonably have of being able to connect at a personal level with plants and animals, rock and wind? Fortunately for their peace of mind, the question seldom arises. The world's comprehensibility and responsiveness are almost always taken for granted. Consider a commonplace scene: a man and his dog. The man reads a newspaper; at his feet the dog wags its tail. What can the dog really know of the man's world of thought, his easy habitation in times past and yet to come; and what can the man, with all his imaginative power, know of the dog's world of odor? What real mutual understanding and communication can there be? Little. An abyss separates the two beings, yet they are often seen as a perfect picture of companionship and compatibility. If this abyss can exist between a man and a dog, which is humankind's oldest domesticated animal, what is one to think of wild animals—the cockroaches, for example, that scuttle across the kitchen table in the dark? Or still further from humans—plants and inanimate nature?[12]

The English diplomat Harold Nicolson noted in a diary entry of 1939, "Nature. Even when someone dies, one is amazed that the poplars should still be standing quite unaware of one's own disaster, so when I walked down to the lake to bathe, I could scarcely believe that the swans were being sincere in their indifference to the Second German War."[13] Here speaks a thoroughly modern and sophisticated man. Disenchantment has proceeded far in the twentieth century, yet how far remains an open question. Modern science and technology indeed encourage a cool and abstract view of nature, but even scientists and technicians, outside the austere milieu of their laboratories, still see (I dare wager) a mountain stream as playfully alive and twinkling stars as possessing some sort of responding intelligence.

Considering the matter historically, one may well wonder at the extent to which the balance has shifted in steady progression toward disenchantment. In the Western world, even as the new scientific view of the late seventeenth and early eighteenth centuries sought to reduce nature to extension and number, Romanticism rose to restore body, color, warmth, and sentience to it. Even as analysis—the idea that nature can best be understood by isolating its components from their natural context—became *the* scientific method, the doctrine of God's providence, with its message of nature's wholeness and integrative purposefulness, was gaining popularity.[14] And what of our own time? Even as departments of botany and zoology fold—that is, even as "plants" and "animals," categories irrevocably charged with human partiality and feeling, disappear, reduced by analytical means to the far more abstract, impersonal units of genetics and molecular biology—ecology comes forward to restore the balance by emphasizing how teeming life-forms and their inanimate environment constitute a system, even a community. Ecology appears to favor connectedness. It is more disposed to search for interdependence than for separateness. And so even as a technical science, ecology can be a balm to modern men and women seeking escape from isolation.

But ecology broadly understood is more than science; it also flourishes as a popular movement—a new faith for Westerners disenchanted with the old Judeo-Christian religion. Two tenets are prominent in the new faith. One is captured by the poetic expression "web of life" and the other by the belief in the existence of an almost human level of sensitivity and consciousness in nonhuman organisms.[15] The second tenet makes it possible for people to feel more emotionally connected to plants and animals. Animals, if not plants, can respond to our overtures in a personal and intelligent way. The web of life, then, is not just the impersonal working out of various kinds of interdependence; it is active—even intentional—cooperation. To a naturalist and lover of nature, if the world out there seems indifferent, we have only our own egotism and obtuseness to blame.[16]

The Mineral Realm

Consider now the mineral realm. Modern astronomy reveals a universe that is almost wholly devoid of life. Even on planet Earth, life exists only as a thin coating. Is not the mineral realm—the intensely hot or cold stuff in outer space as well as the hard rock on Earth that breaks the plowshare—fundamentally indifferent to human projects and well-being, to life itself? The question, though reasonable enough from a modern viewpoint, was hardly ever raised in the past and is indeed seldom raised even in the present. Heavenly bodies were and are as easily anthropomorphized as plants and animals. Even to Copernicus the sun was a sort of seeing eye guiding the orderly procession of the planets, and scientists of our day still anthropomorphize when in mufti. I speak here of the West, which is disposed toward abstraction that sooner or later depersonalizes nature. As for non-Western societies, many show a similar disposition, as, for example, when they create an image of the cosmos anchored on a highly generalized schema of cardinal points, the movement of the stars, and the cycle of seasons that govern the rhythm of agricultural life. What is the consequence of this creative-abstractive effort? Does it encourage depersonalization? The answer is unclear, for in these cosmic schemata, the sun and other stars, though not as vividly personalized as are Mesopotamian and Olympian deities, nevertheless respond to human ritual and prayer. They are not indifferent; they are connected to and have an impact on key human activities. Moreover, the very orderliness of the cosmos implies providential care. The cosmos thus conceived is consoling and promotes confidence in the conduct of life; understandably, societies in widely separate parts of the world have adopted this type of conception.[17] But what is the reality? The reality can be harshly otherwise; just think of violent weather, floods and droughts, tempests at sea, volcanic eruptions, and earthquakes. True, the seasons alternate predictably, but the timing is all too often off. Spring arrives too early or late, summer is too dry or cold, in utter indifference to human needs of survival.

Indifference is even more feared than malevolence. Indifference in nature is so unacceptable that, until the rise of modern science, it is universally suppressed. One can fight malevolent nature with verbal and emotional appeals, or with counterthreats, in a ritual setting. But what can one do with a nature that follows its own paths, deaf to all human appeals and threats? Even today, people find it hard to acknowledge nature's indifference. For example, rather than dismiss as a meteorological fluke the downpour that washes out a football match, the fans react in frustration and resentment as though it were a deliberate spoiler. Nature cares—perversely! The flood not only destroys, it seeks to destroy. The movie *Twister*, made in 1996, treats the tornado as both villain that chases and big game that is chased. Even the National Weather Service defers to the anthropomorphizing tradition; their meteorologists do not think it at all odd to call tropical storms Camille, Hugo, Fran, and Andrew.

Nature's sudden and violent turns teach a lesson of indifference. Nature's great permanencies—the "eternal" hills and valleys, forests and deserts—teach the same hard lesson, but because the great permanencies do not disrupt and intrude, the lesson they impart is harder to learn and easier to forget. In the West it has been successfully covered up by residual pagan beliefs and, since the eighteenth century, by Romanticism, whose poets like to think that as they weep, so do the willows, as they brood, so do the mountains. Poets, however, have not always thought this way, have not always indulged in what critics call the "pathetic fallacy." Consider the greatest poet of Western antiquity—Homer. He truly learned the lesson of nature's indifference. His sonorous syllables—"wine-dark sea," "rosy-fingered dawn," "Poseidon, shaker of earth," and so on—have (as C. S. Lewis puts it) "stereotyped the sea, the gods, the morning, or the mountains," and have made it appear that "we are dealing not with poetry about the things, but almost with the things themselves." Homeric pathos is profound and strikes hard because it "comes from the clash between human emotions and the large, indifferent background which the conventional epithets represent."[18]

Overcoming Disconnectedness

Even as I draw attention to disconnectedness and indifference, I show how both can be overcome—how people escape from both. It is difficult to speak of the one without the other, for disconnectedness implies connectedness, indifference implies attentiveness. Culture, as I have repeatedly noted, is the most common device for moving from one state to the other. But people are rarely aware of *this* important function of a cultural performance or product because their attention is directed to more pressing needs and purposes to which that performance or product answers. For example, good harvests are a pressing need. A ritual is "invented" and performed to ensure good harvests. That is its most public reason for being. Less public and official is the ritual's aesthetic potency—its existence as a work of art that engages and elevates a people's aesthetic sensibilities. Largely unacknowledged, except in critical social-science literature, is the power of ritual to legitimize the orders of society. Hidden almost wholly is ritual's efficacy in overcoming discreteness—in connecting and fusing individual and distinctive entities into larger wholes.

Another, more prosaic example from everyday contemporary life helps to enforce the point. Consider the suburban house. A proud owner, when she thinks of her house at all, is likely to think of its architectural good points, its prestige and value. The house's basic function as shelter is taken for granted. Even more taken for granted is its hidden power to integrate people and their activities. The hidden powers of human artifacts tend to remain hidden unless, for some reason, they are brought into the open. Why anyone would want to do so is an interesting question, for the consequence of such unveiling is almost always ambivalent, the gain in knowledge not fully compensating the loss of innocence.

Touch and Communal Singing

Ways to escape the self are as numerous as sociocultural practice. One broad practice is "bodily contact." It is the most direct method and is as much biological-instinctive as cultural. Bodily contact was more common in premodern times, more common among people with little material means than among affluent people.[19] In a hunting-gathering band, huddling, fondling, and caressing occur frequently among adults, and not just between adults and children. Young men, in particular, like to sleep together in clusters, with arms and legs slung over one another's bodies as though they were bands of lovers. Bodily contact establishes a feeling of oneness so strong that it can transcend even close kinship ties.[20] Communal singing has a similar effect, according to the musicologist Victor Zuckerkandl. In preliterate and folk communities, people gather to sing and so make a reassuring bubble of sound around themselves.[21] There are singers, but no listeners—no outsiders to evaluate the performance and so make the singers self-conscious. Because looking tends to create distance, eyes may be closed in communal singing to enhance the sensation of blissful immersion in sound.[22]

If human togetherness were the sole aim, tones without words should suffice. People, however, feel a strong need to be emotionally engaged not only with other human beings but also with the nonhuman world of plants and animals, rock and wind. For this to occur, says Zuckerkandl, words must be used. Words alone, as in ordinary speech, can capture things for people, making them a part of their world. But the capture is insecure because it is insufficiently felt. When words are sung rather than merely spoken, people and things are finally able to resonate emotionally; the separateness is then fully bridged.[23]

Group Activities

All sorts of group activities have the power to repress the self, especially when these require coordinated movements, as, for example, farmers working in the field, soldiers marching to the music of a military band, or people enacting their designated roles in a ceremony. An awareness of the Other—an indifferent or hostile reality out there—further intensifies group solidarity and weakens the feeling of individual separateness. To farmers laboring together, the fields to be plowed and the weeds to be uprooted constitute the Other. To soldiers, the Other is the sharply defined human enemy. But it is always present, hazily or vividly, for any group that is engaged in a common activity and shares a way of life. A possible exception is cosmic ritual, which in principle is all-inclusive. But there too, something lies outside: Chaos, which the cosmic ritual is designed to forestall or tame.

War is the bane of humankind. Like other group activities, it offers the psychological reward of belongingness. But it does so more effectively than most, which is one reason why it perdures. Military training provides all the right ingredients for escaping the stresses of individuality. Already mentioned are the role of the implacable Other, the calculated submersion of self in lockstep marching. To these one may add such standard practices as the wearing of a uniform, the inculcation of instant and unthinking obedience to commands, the shouting or chanting of slogans in unison, and rites of humiliation. Under military discipline, the self that feels insignificant and frail, the self that is more bewildered than pleased when confronted by real choice, finds salvation in group freedom and power—ultimately, the freedom and power to destroy and kill.[24] Other psychological rewards include a sense of being on the right side and, hence, being in the right. A little-noted effect of this moral posture is that it allows soldiers to retain, individually, a sense of personhood and dignity even while they are immersed in the larger whole. Other agonistic groups such as athletes contesting in the sports arena and political protesters pounding the pavement to a confrontation enjoy similar rewards. There is, however, an important difference. Soldiers in battle are permitted to destroy their enemy utterly. Total destruction of the hostile Other—enemy soldiers and their country—confirms the winning side in its belief that the Other is undifferentiated evil, conspicuously lacking in any individualized human quality; for without doubt, all buildings look alike as rubble, all human beings look alike as corpses. Finally and perhaps most sinister of all, war legitimizes the practice of disemboweling or exploding another person.[25] What makes another person unique beyond surface appearance and public behavior is the possession of an inner self that is frustratingly inaccessible even to intimate outsiders. That inner self—its store of feelings and thought—is "opened out" in violence.

The Built Environment and Oneness

Culture can be primarily an activity, as in etiquette, ritual, games of war. Culture is also the end product of an activity (skill)—a material artifact. Whereas we readily see that certain kinds of activity promote group cohesion, we less readily see the artifact as providing the same service. Consider the built environment. At the scale of a room, the effects on human behavior and solidarity are so omnipresent and subtle that people do not—or hardly ever—pause to marvel.[26] In the family room, family members almost always do different things, live in separate, strikingly dissimilar worlds of awareness: the baby crawls on the floor, the teenager studies Latin, the mother balances the budget, the father dozes before the television set. Yet they do not feel this separateness. On the contrary, they feel themselves to be a close-knit family, and any observer of the scene would conclude the same. The enclosed space of the room, a cheerfully illuminated interior set against the darkness outside, encourages a sense of oneness. Likewise, the pictures on the wall, the coordinated pieces of furniture, all attest to a whole that is greater than the sum of its parts. In this model modern family, there is little bodily contact. Perhaps little is needed because the constructed space and its furnishings already do much to convey a feeling of cohesion.

Take, as another commonplace example, the classroom. It differs from the family room in that there, all students are presumably engaged in the same activity, namely, attending a lecture. They may feel themselves "one"—one student body—for that reason. The professor also tends to see the students as one—in any case, as much alike—and in this perception he or she is undoubtedly influenced by the physical environment. In the classroom all the chairs, neatly arranged in rows, are identical, which promotes the illusion that the students who sit in them are alike—that they all have much the same body shape and weight, much the same sensory equipment, much the same kind of mind and intellectual preparedness, absorbing professorial wisdom in much the same way. What a shock when the professor reads the blue books!

And beyond the scale of a room? What can we say about the impact of house, neighborhood, and town or city on the feeling of oneness? I would say that the impact is certainly there, but it is more intricately layered and more a product of visual perception and conscious awareness. A house—say, a modern middle-class home—is made up of rooms, each of which may be designed for a different kind of activity. Such a house not only promotes separate spheres of life but also imparts greater coherence to each sphere, if only because people within it are aware of other kinds of activity in other rooms. The house itself is an architectural whole, evident especially when viewed from outside; its perceptual integrity is a subtle yet powerful reminder to inmates that they are not isolated beings but members of a group. The entire house, both as haven and as symbol, sets up a polarity of inside and outside, "us" and "them", and this has the effect of reducing the differences—the clashing heterogeneity—within each pole. The larger environments of neighborhood and town work in a similar way. A neighborhood's architectural distinctiveness has in itself an effect on group identity. Through day-to-day visual experience the inhabitants of a neighborhood know when and where they have crossed the line from a region that is "us" to a region that is "them." As for the town, even one that lacks architectural character still stands out, literally, from its surrounding fields. Any sharply circumscribed town can thus be a unifying landmark for its inhabitants. It performs this role by simply being there for everyone to see and experience. A town favored with architectural monuments enjoys the added advantage of symbolic resonance—a resonance that is further heightened when ceremonies are conducted around them and stories are told about them.[27]

The word "ceremony" takes us back to the idea of culture as gesture and activity; it points to worlds evoked by motions of the body rather than by the making of buildings and things. By motions of the body I mean everything from dining etiquette to rites of agriculture and of war. I discussed these gestures and activities earlier. What I have yet to address with any fullness is storytelling. Storytelling is of prime importance because language is at the core of human culture. Without language, the uniquely human ways of transforming, covering up, and escaping are inconceivable.[28]

Speech and Bonding

Speech binds. Human beings know for sure that they live in the same world when they apply the same words to the same things—when they speak alike. If I am not a botanist, why do I still want to know the name of a flower? What additional information do I gain when I am told that I am looking at an African violet, a specimen of *Saintpaulia ionantha?* None. Knowing its name reassures me not because I know more about a plant, but because I now share one more term with other people, which gives me the impression that I share one more bit of the world with them. For a people to sustain the belief that they live in a common world, their conversational vocabulary has to be limited, as it almost invariably is.[29] In modern society, even when men and women chatter away, they seldom venture beyond a small set of different words—not much more than a hundred—in the course of a day. Bonding among members of a group is further strengthened if they develop a distinctive pronunciation, a jargon; and linguists assert that every close-knit human group has its own manner of speech that sets it off from others. Indeed, when we ask, What aspect of culture most sharply differentiates one group from another? the answer is, Not food, dress, housing, kinfolk network, and suchlike, but language. The number of distinct languages in the world far exceeds the number of distinct lifeways. Two tribes living in the same geographical region and sharing many cultural traits may each have its own language, making verbal communication—the uniquely human way of bridging isolation—nearly impossible. Although the number of languages in the world is rapidly diminishing, even in the mid-twentieth century some three to five thousand were still in use, and linguists believe that in earlier times many more were spoken. I am led to fantasize a distant past when every knot of people had its own special way of speaking, jealously guarded against contamination from outside. Being able to communicate with another group enlarges the scope of experience and so better enables a people to survive in its own periodically threatening natural environment. But in the past perhaps an even more pressing need was internal cohesion—a feeling of oneness and sameness that depended on the existence of a closed world, beyond which lay an incomprehensible Other.[30]

I have said "in the past," but even as small local languages are dying, vigorous movements to revive some of them have arisen in recent decades in various parts of the world, together with a resurgence in regional and nationalist consciousness. The desire to retain or resuscitate one's unique cultural signature in an increasingly undifferentiated cosmopolitan (global) society has translated into surprisingly effective political programs. Localism or particularism is now widely and explicitly recognized as a good—an attainable political goal. Not so well recognized—indeed, repressed in the process—is an important side benefit of the movement, namely, escape from the openness of the world in which a person stands alone, charged with the bewildering task of defining himself, to the haven of language, in which an individual's sense of a self is confirmed and stabilized each time he speaks, knowing subliminally as he does so that when he speaks he exercises the power to include and exclude. A further benefit is this: Although the identity thus welded is group rather than individual, within the group an individual can always build up a particular sense of self by developing, say, a mannerism or a special skill. What makes living in such a world so reassuring is that whenever that specialness—never in any case egregious—becomes isolating and uncomfortable, the individual can quietly lay it aside and slip back into the warm bath of common speech.

Native-tongue patriots may say that a language is to be protected or resurrected not so much because it provides a social glue or warm communal bath for those who need to escape chilling selfhood, but because it is poetry—a unique and untranslatable way of seeing and knowing the universe. Maybe it is. Any threatened language should be given the benefit of a doubt—that is, treated as though it might be an inspired and irreplaceable product of mind and emotion. Still, I am inclined to guard against excessive romanticism, for on the face of it a nation of poets—a people whose speech is regularly punctuated by words of freshly minted significance—is unlikely. Ethnographic evidence and our own experience among speakers of a foreign tongue hardly confirm linguistic creativity as a common state of affairs. A well-meaning outsider may think a foreign tongue poetry because many figures of speech in it are new to her. But they are not new—they have probably become verbal tics—to native speakers. It is both humbling and healthy to remember that when people anywhere, including erudite academics at the campus club, meet to talk socially, they are seldom poetic whatever language they use. They are more likely to say, "Pass the salt," "How are the kids?" and "So-and-so didn't get tenure" than anything that stirs the emotions, elevates the mind and the spirit.

Speech not only binds human individuals but also binds human individuals and groups with their nonhuman environment. It does so most effortlessly and effectively by means of similes and metaphors, which are a universal feature of language. Apparently human beings can only know who they are through the use of animal and plant references: "I am a fox; you are a pig; he is a prickly cactus; she is a lotus blossom." In the process of learning who they are, people also become aware of their intimate ties to other living things; the two processes are inseparable, melded into one by the character of language.[31] As for things in the mineral realm, anatomical metaphors such as foothills and headlands, the spine of a ridge, the mouth of a river, the face of a cliff, make them all seem familiar and personal. Indeed, language tricks people into believing that rises and hollows, wind and rivers, are all in some sense alive. And because human beings and human speech are coeval, there never was a time when speech did not generate this useful and reassuring illusion. Language animates; that and human bonding are two of its most primitive and potent effects. Historically, the problem that confronts human beings is not how life has emerged in a lifeless universe, but rather (as Hans Jonas puts it) how a warm body can turn into a cold corpse. Ordinary speech lacks neutral ways of referring to that which is not alive. The inanimate is a sophisticated idea that depends on the prior conception of the animate.[32]

We think we are losing the knack of figurative utterance. We speak prose, and pundits have offered the opinion that this prose is becoming more and more impersonal. There is, however, an important difference between impersonal prose and plain prose. As the Royal Society of London recognized in its rationalist heyday, speaking or even writing plain logical prose with minimum metaphorical embellishment is a high and rare achievement. In any age, it lies well beyond most people's competence. Other than those occasions when officialese takes over, we modern people, no less than people in earlier times, are "poets." We speak poetry—just not good poetry; and although fresh metaphors are normally beyond our grasp, our daily speech is nonetheless peppered with metaphors, other tropes, and literary conceits. "You pig," "hi, honey," "angry sky," "bullish market," and suchlike are certainly not original, yet so long as such expressions are used we maintain a personal link between nature and self. No wonder the idea of the world's indifference, other than under some dire circumstance, seems so oddly contrary to experience and common sense.

Speech and Solitude

Through most of human history, speech has helped to cement group cohesion. In numerous myths and lores, the kinship of all things—some kind of mutual understanding at the level of feeling and verbal communication—is assumed. Speech nevertheless is not just social glue. It is also an instrument for critical reflection. Used in a certain way, speech empowers people to penetrate its own social character, to put into question its propensity to elevate the fetishes of the group into nature's norm, its power to mask the world's disjunctions and indifference. A sharply critical or scientific manner of speaking eschews the rhetorical devices that so effortlessly rope the nonhuman into the human world. In the interest of truth, it sets aside human bonding needs. Ironically, such a language—say, that of chemistry—creates its own bonding among users. The perhaps chilling distance that chemists put between themselves and the nature they study is compensated by the unusual closeness with which they work with one another. By speaking in a way understandable only to themselves, chemists can forget their richly faceted, sometimes conflicting, personalities. They assume the much simpler role of experts, and as such they possess slightly differing points of view, which they offer to colleagues in a warm bubble of mutual appreciation. Life in this specialized community can be fully satisfying unless or until some other dimension of reality rudely intrudes.

In a well-coordinated and well-designed room it can seem as though the armchair and its ottoman, a standing lamp and the adjoining side cabinet, are "conversing" with one another, such that a polite person would hesitate to disrupt the exchange by passing between them. When human beings stand side by side talking, is there a similar sense of coordination, of souls in deep and sympathetic converse that should not be lightly disrupted? I would say yes, but even more often no, at least on social occasions, which means on most occasions. The fact is, people seldom truly speak with or listen to one another; more often than they care to admit, they deliver soliloquies, with each individual using another's remark merely as a launching pad for his or her own performance. It is tempting to say, "Well, in the old days, friends and family had genuine conversation as they sat around the fireplace. In our time, with so much television watching, we have lost the art." Tolstoy would have disagreed. Golden days in talk, as in so much else, are a myth. "I don't know how people were in the old days," he wrote early in his literary career, "but conversation there can never be. . . . It is not from any deficiency of intelligence but from egotism that conversation fails. Everyone wants to talk about himself or about what interests him."[33] I am reminded of André Gide, who reportedly believed that people were not really interested in what he had to say and that hence he had a tendency, even in his written work, to rush the ending. He based this conclusion on a common experience. Interrupted by a telephone call in the middle of a story he was telling, he was never asked, when the call ended, "And what comes next?"[34]

Egotism may indeed be the heart of the problem. Egotism, however, is not just the moral defect of some people. It is a universal characteristic rooted, in part, in a human individual's uniqueness of body and mind, uniqueness of experience. Within any human group, significant differences exist; and though these may be covered up by the wiles of culture, they remain a recurrent source of conflict and frustration. Individual selves, repeatedly thwarted in major and minor ways, become highly sensitized to their own needful existence. They ask plaintively, though usually only to themselves, "Why isn't anyone paying attention? Why am I not deferred to?"

Human beings, moreover, have projects, which are a fertile ground for ego inflation. By projects I mean not only those of large scope pursued by ambitious individuals, but also the numerous purposeful undertakings of the day known to everyone. Consider again André Gide's sad but wise awareness that nobody really wanted to listen to him. Listening makes a demand on selflessness. I must give up my thoughts—my project—in order to attend to another's. And this is not nearly as easy to do as we may think. In Simone Martini's altarpiece *The Annunciation*, the Angel of the Lord speaks; Mary listens but discreetly keeps her place in the book she was reading when the Angel arrived. The Angel's mission is no doubt momentous, but it is not hers; she listens with a hint of reluctance, because she was otherwise engaged. In Anton

Chekhov's play *The Three Sisters*, anguished cries from the heart are deflected by a rebuke to the servant or an asinine comment on the weather. Misunderstanding in a familiar setting—in the family and among friends—is a common theme in modern literature: a happy face encounters a face of uncomprehending gloom; things are said but are not understood; distress signals are sent out but are lost in indifference. Human beings have not become more egotistical, only more aware of their egotism—of their disinclination to see or listen, thanks in part to works like those of Chekhov, Tolstoy, and numerous other writers.[35]

If a small vocabulary and the frequent use of clichés promote understanding and communal solidarity, the achievement of verbal-intellectual sophistication can have the opposite effect. The more people know and the more subtle they are at expressing what they know, the fewer listeners there will be and the more isolated individuals will feel, not only at large but also among colleagues and co-workers. Let me use an architectural metaphor to show how this can come about in academic life. Graduate students live in sparsely furnished rooms but share a house—the intellectual house of Marx, Gramschi, Foucault, or whoever the favored thinker happens to be.[36] A warm sense of community prevails as the students encounter one another in the hallway and speak a common language, with passwords such as "capital formation," "hegemony," and "the theater of power" to establish firmly their corporate membership. Time passes. As the students mature intellectually, they move from the shared life of a house to rented apartments scattered throughout the same neighborhood. The apartments are close enough that friends still feel free to drop in for visits, and when they do the entire living space is filled with talk and laughter, recapturing as in younger days not only the bonhomie but also the tendency to embrace wholeheartedly the currently headlined doctrine. Eventually the students become professors themselves. They begin modestly to build their own houses of intellect and add to the structures as they prosper. Because each house bears witness to a scholar's achievement, it can be a source of great personal satisfaction. But the downside is, who will want to visit? And if a colleague or friend does, why should she spend time in more than one room?

Social scientists claim that a tenement building where people hang out the washing or sit on the stoop to socialize can be a warmly communal place. By contrast, a suburb with freestanding houses is cold and unfriendly. I am saying that the same may be true of intellectual life as one moves to larger houses of one's own design. Both types of move—socioeconomic and intellectual—signify success, and with both the cost to the mover can be an exacerbated feeling of isolation.

Chaos and Order

Although cultural covers are normally effective, they are breached from time to time by reflective individuals, of whom almost every society has a few. The buried dilemmas of disconnectedness and indifference—the chaotic character of reality as such—then rise to the surface. The mind turns out to be a double-edged tool; if as constructive intelligence it produces covers, as critical intelligence it destroys them. Many of us, I believe, strive to maintain a cover even though we suspect that it is little more than that. The following account drawn from the ethnographic literature is revealing and not all that unusual. A Navajo father instructs his children in the use of a string game, showing how it connects human life to the constellations, as prelude to telling the Coyote tale. The game and the tale may be entertaining in themselves, but they have a deeper purpose, which the father explains thus: "We need to have ways of thinking, of keeping things stable, healthy, beautiful. We try for a long life, but lots of things can happen to us. So we keep our thinking in order by these figures and we keep our lives in order with these stories. We have to relate our lives to stars and the sun, the animals and all of nature, or else we will go crazy or get sick."[37]

The Navajo father commends thinking for its power to produce temporary stays against disorder. Many societies, however, recognize that thinking without some immediate, practical end in mind can cause unhappiness and that, indeed, it is itself evidence of unhappiness. Happy people have no reason to think; they live rather than question living. To Inuits, thinking signifies either craziness or the strength to have independent views. Both qualities are antisocial and to be deplored. One Inuit woman was overheard to say in a righteous tone, "I never think." Another woman complained of a third woman because she was trying to make her think and thus shorten her life.[38] Even in modern America, thinking is suspect. It is something done by the idly curious or by discontented people; it is subversive of established values; it undermines communal coherence and promotes individualism. There is an element of truth in all these accusations. In an Updike novel, a working-class father thinks about his son reading. It makes him feel cut off from his son. "He doesn't know why it makes him nervous to see the kid read. Like he's plotting something. They say you should encourage it, reading, but they never say why."[39]

I have chosen the words "isolation" and "indifference" to capture a fundamental human experience. Other people may choose other words and concepts. One of the most common is "chaos," which invokes the ultimate in disconnectedness, isolation, and indifference. Even as the Navajo father uses a string game to show his children how human fate is tied to the constellations, he feels and fears the undertow of chaos. Likewise Aua, an Iglulik Eskimo. When the Danish explorer Knud Rasmussen tries to get Aua to articulate a coherent philosophy, Aua replies that it cannot be done; moreover, it is presumptuous to do so, as presumptuous and futile as to build elaborate material shelters and camps.

Whatever people may say and would like to believe, they know through the shocks of experience that nature is indifferent and can be chaotic. "To hunt well and live happily," Aua notes, "man must have calm weather. Why this constant succession of blizzards?" And "Why must people be ill and suffer pain?" Personal misfortune seems quite unrelated to good or bad behavior. "Here is this old sister of mine: as far as anyone can see, she has done no evil; she has lived through a long life and given birth to healthy children, and now she must suffer before her days end. Why? Why?" "You see," says Aua to Rasmussen, "you are equally unable to give any reason when we ask you why life is as it is. And so it must be." In a world so full of uncertainty, the Iglulik seek comfort in the rules they have inherited. To quote Aua again—and what he says is remarkably similar to the Navajo father's view—"We do not know how, we cannot say why, but we keep those rules in order that we may live untroubled."[40]

Reflective individuals may be defined as those who suspect that the order and meaning they discern and strive hard to maintain are little more than a measure of their desperation. Aua was such an individual. Another, also an Iglulik, was Oqutagu. He aspired to become a shaman but changed his mind during the period of training. To his kinsmen he said he was not good enough; to the friendly outsider Knud Rasmussen he explained that the real reason was that he had come to doubt his master's claims to reading the signs of nature and to establishing contact with helpful spirits. He saw such claims as lies and humbug—well-meaning, perhaps—manufactured to provide reassurance to a timid people.[41] The anthropologist Monica Wilson asked the women of an African village why they set such store by their ceremonies. Was it because they exerted real power on the external world? Their answer was always the same: Such ceremonies were conducted for inward rather than outward effect; they served "to stop people going mad."[42] I am reminded of W. H. Auden's gloomy poem "Death's

Echo," in which he says, echoing the ancient Greeks, that not to be born may well be the best, but there is always a second best, which is formal order, the "dance's pattern." To make some sense of life—to prevent ourselves from going mad—we have one ready means of escape, and that is to dance while we can.[43]

"Unironed handkerchiefs could lead to madness," writes Iris Murdoch.[44] People must opt for order somewhere down the line. A touching confession of helplessness before the world's bewildering complexity comes from the distinguished anthropologist Claude Lévi-Strauss. He has been accused of reductionism, of suggesting that structural analysis has the power to illuminate human experience and social reality. Lévi-Strauss denies this as "outrageous." The possibility, he says, has "never occurred to me. On the contrary, it seems to me that social life and the empirical reality surrounding it unfold mostly at random." As Lévi-Strauss picturesquely puts it, "disorder reigns" in social life's "vast empirical stew." He, for his part, chooses to study only its "scattered small islands of organization." Moreover, these "islands" refer not to "what people do, but [to] what they believe or say must be done."[45]

People try to maintain orderly material and conceptual worlds, but if a choice has to be made between them, no doubt the conceptual will be selected as the more important to keep up. Even when material houses collapse, life can still more or less continue in temporary shelters; in any case, they can always be rebuilt when resources are once again available. This is not the case with an edifice of beliefs and ideas. Its decay undermines morale even more than the fall of physical structures; and once down, it is much harder to rebuild. Reflective individuals such as Aua, Oqutagu, and the Navajo father, by raising fundamental doubt on the nature of things, threaten the conceptual edifice. Understandably, all societies try to silence such individuals. One possible exception is the modern West. In modern Western society, questioning basic beliefs and customs is not simply an odd happening initiated by a bold and gifted individual in exceptional circumstances—such as when he or she is interviewed by a sympathetic outsider—but fairly common, emerging from an intellectual climate that has, at least ostensibly, the approval of society itself.

A strongly analytical and critical disposition of mind, sustained over time, can lead to cynicism and despair. In the West this has not yet happened to a pronounced degree, and one reason is ironic: The same hard questioning that has corroded traditional covers has enabled Westerners to build a new one—the dazzling technological world that has its own great powers to shield, entertain, and distract. Still, in the course of the last two centuries, critical thinking has undoubtedly dented the modern person's sense of what the world is really like, raised disturbing questions about the true character of human relationships, including those long considered sacrosanct, as well as the true character of the relationships between, on the one hand, human beings, and, on the other, animals, plants, rock, the vast silent space. Would not these foundation-shaking queries be another reason, maybe the deepest one, for the vehemence with which the West is sometimes attacked? Besides its egregious faults of imperialism, racism, and speciesism that are generic to civilization, the West has allowed a way of thought to develop that is uniquely destructive of cultural covers and escape routes, not only other people's but also its own.

Landscape and Human Separateness

To speak of culture as cover or covering up is to give it a negative meaning. But this is an unconscionably partial and distorting view. Culture exposes as well as masks; speech, besides obfuscating, also clarifies. Moreover, culture introduces new things into the world: artifacts that are useful and artworks that can be not only beautiful but true. Occasionally it produces objects that embody human excellence, admirable in themselves and not just because they enable people to survive, hide what they do not wish to see, or even correctly picture—as in a scientific law—an aspect of nonhuman reality. An example of an artwork that enlivens the senses and teaches people to look at and think about the world in a new way is landscape painting. Of course, other kinds of artworks may show similar virtues and powers. If I take up landscape painting at this point rather than, say, portraiture, woodwork, or sculpture, it is because landscape painting happens to support a principal thesis of the present chapter, namely, that human beings remain separate even in a world that seems otherwise connected and cogent.

Landscape painting requires, as a necessary condition, a society that is prosperous and exercises a large measure of control over nature.[46] In China, it was to become a distinctive genre during the T'ang dynasty and to reach a peak of excellence during the Sung dynasty, both being periods of rapid advance not only in the arts and technology but also in interregional trade. In Europe, landscape painting first emerged during Greco-Roman antiquity, disappeared in the Dark Ages, and reemerged around 1400—that is, during the Renaissance.[47] But haven't people always found it pleasing to stand on an eminence to look at a composition of hills and valleys, woods and meadows? Apparently not. Aesthetic appreciation of a panoramic scene appears to be an acquired taste—a rare taste. Rarer still is the desire to capture the scene in a work of art. Rarest of all is the desire to depict accurately what the eyes see, as distinct from what the mind knows. Only Europeans have wanted it badly enough to spend some four centuries perfecting the art, which they also considered a science, for the realism they aimed at—the conformity to actual visual experience—could only be achieved through a sophisticated understanding of perspective, the properties of color under different lighting, and the actual shape and structure of (say) mountains, as well as how such shape and structure would change at various intervals of distance.[48]

A landscape is, above all, a composition. It reveals harmonies large and small, most of which are invisible to the people who live there and must attend to immediate needs. Landscape demonstrates the advantage of distance. Only from a certain distance can an overall structure be discerned and a unique type of relationship, emotional yet somewhat cool, be established between a human individual and reality. But from a distance, harmonies of life and environment are not all that a viewer sees. He or she also sees discontinuities and isolation—the world's indifference. True, not many viewers do so, and assuredly not the tourists who flock to gawk at famous panoramas. But a discerning few can, and among them were the Old Masters. In a famous tribute to them, W. H. Auden notes how a landscape masterpiece may show, as its central theme, adults waiting reverently for the miraculous birth of Christ, but elsewhere, as a sort of commentary, children skating on a pond on the edge of the wood who "did not specially want it to happen." In Brueghel's *Icarus*, which illustrates the story of the brash youth falling out of the sky, the poet asks us to note

> how everything turns away
> Quite leisurely from the disaster; the ploughman may
> Have heard the splash, the forsaken cry,
> But for him it was not an important failure.[49]

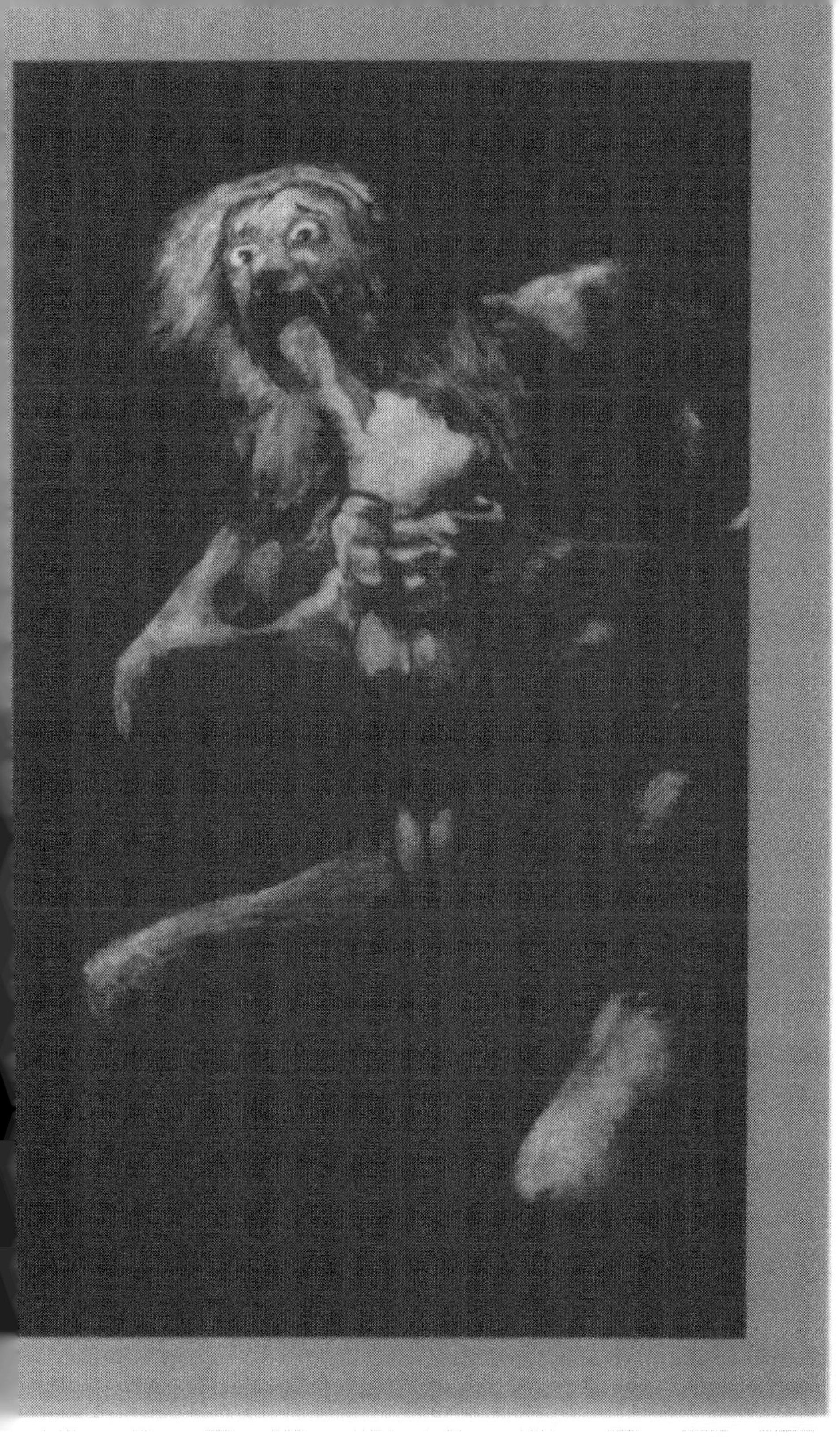

4

HELL

Imagination's Distortions and Limitations

• • •

Culture is the product of imagination. Whatever we do or make, beyond the instinctual and the routine, is preceded by the kernel of an idea or image. Imagination is our unique way of escaping. Escaping to what and where? To something called "good"—a better life and a better place. "Good," for most humans historically and for many even today, means physical survival and a little extra. From the need to ensure both comes the desire for tangible things: green pastures, rich harvests, sturdy shelters, possessions, many children, and so on. Good thus translates into goods; so much of life turns out to be a struggle not for good but for goods.

Even in the midst of material abundance, an isolated individual is vulnerable. Security lies in mutual help and, psychologically, in the possibility of losing self and its anxieties in the larger whole—in being inconspicuous. On the other hand, the opposite and more risky path also works. Security is obtained, not by losing oneself in the larger whole, but by wrapping the self in the trappings of power and prestige. Power and prestige being social, seeking them entails active participation in the group, but this time conspicuously—exhibitionistically—with the cunning help of imagination. Good then means far more than survival. Indeed, to the powerful and confident, physical survival as such is taken for granted and displaced to the lower realm of mere biological and animal life. Prestige is measured by how far one can rise above it.

Human vulnerability is not only subjection to physical pain, disablement, and death. It is also a corrosive sense of emptiness at the core of being, which one tries to overcome with the drowsing fumes of alcohol and drugs, and with socially approved work, projects, entertainments, but above all with human company, the hum of small talk that plugs every opening to ominous silence. We need other people, then, not just as a bulwark against external threat but as an effective diversion from having to confront this inner vacuity.

People want the "good" that is real, which is one reason why they latch on to material things. But material things, for all the direct satisfactions they provide, apparently cannot stand alone. To be truly valued, they need the support of imagined worth. The gap between palpable worth and imagined worth, with the latter backed by the power of society, is greatest in the art objects that are the pride of civilization. A city rich in art and architecture can, in a certain light, seem unreal, because such works boldly present themselves as products of a highly imaginative (even fanciful) mind and also because for all their materiality, they are wrapped in prestige, whose literal meaning is "sleight of hand," and glamour, whose literal meaning is "magic."

Beyond good things, people want good human relationships. A deep psychological reward of good human relationships, seldom acknowledged, is that they feel *real;* certainly the love that one bears toward the beloved and the beloved herself are real as no other experience or object in the universe is. Yet it is in human relationships that deception and self-deception are most likely to occur. It is there, as in no other encounter, that imagination can make one plunge into hell or soar to heaven.

The imagination that is the pride of the human species is also at the root of much anxiety and sorrow. Harsh as reality is, too often what people dread is "in the mind" rather than real. In the history of a people and in the life span of an individual, which preponderates—good or evil? Arguably, evil. History can make unbearable reading because it exposes the extent and weight of misery among humbler folk. As to the individual life, even the favorites of the gods, such as Tolstoy or Goethe, claim that they have known few moments of genuine happiness. A fine but harrowing novel can be read over and over again; few, however, would want to relive their life even when, from an objective point of view, it has been good. Yet as Hans Jonas says, life can still be worth living when the ledger shows a preponderance of misery, because sentience and, above all, imagination—an extremely rare and wondrous cosmic happening—are good in themselves.[1]

Imagination can give us heaven. It can also give us hell. First, there is the obvious question, Why on earth would anyone want to use imagination to create hell? The answer is that one doesn't. Most of us don't. It is not a matter of intention. We have this remarkable faculty, we exercise it, and all too often—to our own surprise—we end up in a hellish place or in some flossy unreal "heaven" that is not much better. The mind—or the psyche at a deeper level—suffers from certain distortions and limitations that have the effect of taking us to where we, in our right mind, would not want to go.

Apparitions Galore

Consider this curious fact: Uncertain and full of danger as life already is, the human mind is ready to make it worse by populating its world with monsters, demons, ghosts, and witches that, for all their lack of substantive reality, feel hair-raisingly real. Functionalist explanations offered by social scientists do not quite answer. Some deep psychological impulsion—a "dark force," as old theologians might put it—seems to be also at work. Let us begin with young children, whose inordinate fears are a hint of what lies ahead. Even children who have led a sheltered life in the care of loving parents can know night terrors that drench them in sweat. What is it they see? What occasional bad experience in a child's life can match the sheer terror of these nighttime monsters? Children, as their imagination develops, learn to fear the dark. Ironically, new mental powers take them to a world that is more rather than less frightening. As they continue to gain strength and self-confidence, however, the phantasmagorical world does eventually recede. Likewise, as a society gains more confidence and lives increasingly by the light of practical reason, its population of demons diminishes. In modern times we tend to forget how they once haunted people all over the world. What may still lurk under the bed today for an overwrought child lurked threateningly not long ago—in times of stress—almost everywhere for everyone.

A major difference between our time and the past, between progressive and static societies, is in the degree of enchantment and hauntedness. "Enchantment" evokes pleasing images, and we may regret their loss, forgetting that good fairies and guardian angels do not exist alone; they have their dark cousins. Here is an illustration of the sort of change that has come about. We think of today's world as swarming with people and of earlier times as far less crowded. In the European Middle Ages, towns might be densely packed, but open spaces could always be found in their midst, and just beyond lay greensward and forest. Nevertheless, medieval men and women felt the constraint of overcrowding. Indeed, whether in the city or in the countryside, a person could hardly stir without bumping into someone. Whom? Not, perhaps, another mortal, but a spirit or demon. Common belief had it that no less than a tenth and possibly as many as a third of all the hosts of heaven fell with Satan. Devils by the dozen were available to torment every man, woman, and child.[2]

The heated imagination that only yesterday gave us a haunted world may still do so in some dark age of the future. Reason and common sense are not a certain bulwark against lurid images, triggered by fear, that surge from primitive recesses of the mind. Childishly, when such images do not come unasked, we miss them and want to revive them for the thrill they provide. As a world made too dully safe by technology drives some people to the visceral excitement of the roller coaster, so modern life's pedestrian and even comely landscapes call for, as needed contrast, pockets of uncanniness and dread.

Wanton Destructiveness

All living things are destructive, if only because to live they have to assimilate other bodies to themselves. Humans, however, show an almost wanton capacity to smash and crush that is tied to their imagination and their ability to make: eggs broken to make omelettes, forests cleared to make books and houses, earth disemboweled to raise monuments of landscape architecture, and so on. But destructiveness as such has an appeal to humans that is not evident in other animals. What is the nature of the appeal? Power. Power is the beginning of an answer. For many people destruction is the clearest evidence of their ability to change the world and hence the most convincing proof of their own existence—their own reality and worth. "[Knocking] a thing down . . . is a deep delight to the blood," says the suave and gentle philosopher George Santayana.[3] Parents will agree as they watch an early, proud accomplishment of their infant, which is to knock something down. Children old enough to make things retain a fondness for destruction. They may even build for that purpose. At the beach they make "elaborate reservoirs of sand, fill them with water, and then poke a little hole in one of the walls for the pleasure of watching the water sweeping them away."[4] To Wilhelm von Humboldt, a distinguished humanist and educator, the sight of a force that nothing could resist had always had great appeal. "I don't care," he wrote, "if I myself or my best and dearest joys get drawn into its whirlpool. When I was a child—I remember it clearly—I saw a coach rolling through a crowded street, pedestrians scattering right and left, and the coach unconcerned, not diminishing its speed."[5] Note Humboldt's admission that he didn't care who were the victims. But, then, haven't there been times when almost anyone might—to shouts of "Let everything go!"—tear down a fence, junk the family furniture, and throw old fond letters into the roaring flames? When a brush fire destroyed not only his California house but also its irreplaceable content of correspondence and manuscripts, Aldous Huxley remarked nonchalantly that it was only "the great denudation" a little ahead of time. In war this strange acceptance, if not appreciation, of destructiveness becomes strikingly manifest. "Theater of war" itself suggests that when one's own life is not at risk—and even when it is—the battlefield can be an exhilarating spectacle.[6]

Cruelty and Limited Imagination

Cruelty is not one of the Seven Deadly Sins of medieval theology, an omission that surprises modern sensibility, for we have come to see the deliberate infliction of pain as possibly the worst evil. "Cruel" and "crude" have the same root; both speak of a rawness that is part of our biological nature, which can be removed through acts of cumulative refinement.[7] Cruelty may thus simply be the effect of an immature mind. Young children are often cruel. As a child, I never hesitated to impale a live worm on the sharp point of a fishhook, something I would rather not do now. But is "cruelty" the right word? The child I was then did not intend cruelty; I just never thought of the wriggly thing in my hand as anything other than bait to be used for a venture that engaged me wholly: fishing. Elizabeth Marshall Thomas provides a more lurid example of thoughtless cruelty. In a much acclaimed book called *The Harmless People*, Thomas presents a charming picture of a Bushman family in Southwest Africa that seems to lend support to the title, but that demonstrates how far primitive life—indeed, any human life—is from harmlessness. Around a fire at camp, Gai the hunter gave his baby son Nhwakwe a tortoise and offered to roast it for him. An old woman, Twikwe, helped. She "held the tortoise on its back, but the tortoise urinated brown urine and Twikwe let it stand up. It stood looking at the flames, blinking its hard black eyes, then started to walk away. But Twikwe caught it again and held it, idly turning it over and over while she talked with Gai about other things." After talking a while, "Gai took the tortoise from Twikwe and laid it on its back." He then applied a burning stick against the tortoise's belly. "The tortoise kicked violently and jerked its head, urinating profuse amounts of the brown urine which ran over Gai's hand, but the heat had its effect, the two hard, central plates on the shell of the belly peeled back, and Gai thrust his hand inside." He pulled out the heart, "which was still beating, and flipped it onto the ground, where it jerked violently for a moment, almost jumping, then relaxed to a more spasmodic beating, all by itself and dusty, now ignored." Gai pinched the gall away from the liver but "left the liver and the surrounding fat because he meant that to be eaten." Let me quote the next paragraph in full:

> The baby, Nhwakwe, who owned the tortoise, came to sit by his father, leaning on Gai's leg and watching, looking smiling into the belly of the tortoise. A tortoise is such a slow tough creature that its body can function although its heart is gone. Nhwakwe put his wrists to his forehead to imitate in a most charming manner the way in which the tortoise was trying to hide. Nhwakwe looked just like the tortoise.[8]

I offer this example and my own childhood treatment of the earthworm not for their shock value but for their ordinariness—

their daily occurrence. Such acts issue, it would seem, from a deficiency rather than an excess of imagination. As a child I simply lacked the power to empathize with the earthworm. The same might be said of the Bushmen as they playfully [illegible] tortoise. But was this really the case? My examples might illustrate the expedient repression of imaginative power rather than its lack. For humans are well able to identify with animals. Note how baby Nhwakwe cleverly imitated the tortoise—saw himself as one—as it struggled to escape from its torturers.

The tough question that such commonplace acts raise is, How would human life be possible if we were unable to inflict pain unthinkingly? We have to eat, after all, and food preparation—the whole sequence from trapping, rounding up, and killing the animal to its skinning, dismemberment, and cooking (roasting, boiling, frying)—entails violence. A disturbing thought is this: May not the ease with which we can, between bouts of laughter and chatting, cut open a turkey and pluck out its innards, prepare us to commit comparable outrages against human beings—those we consider less than fully human—when circumstances permit?

Cruelty and Capricious Imagination

The cat plays with the mouse before killing it. That may be what we say, though in a more scientific state of mind we refrain from attributing cruelty to animals because we question that they have the necessary reach of imagination. Human beings, even young ones, are another matter. How innocence in the human young can combine with satanic inventiveness is shown in Joseph Czapski's story of Russian children splashing water on the corpses of German soldiers they found in the snow so that next morning they could use the frozen bodies as sledges.[9] To act cruelly, the tormentor must crush one type of imagination, sympathy, and exercise another, creative malevolence. John Updike notes that as a child he tortured his toys, "talking aloud to them, fascinating and horrifying [himself]." The toys were a Donald Duck, a Mickey Mouse, and a Ferdinand the bull, a "proto-flower-child" with a "drugged smile." Updike recalls, "I would line them up, these smiling anthropoid animals, and bowl a softball at them as in ten-pins, knocking them down again and again, and all the while taunting them in my mind, like some Nazi interrogating Resistance prisoners."[10]

Sticking a worm on the fishhook is one thing; pulling the legs off a grasshopper, as children may do, is quite another. Of the first, one is justified in saying that it helps the young to acquire the skills of obtaining food; the need to eat must be accepted as a plausible excuse for violence, else one goes mad. But dismembering a grasshopper serves no other purpose than the pleasure of absolute power, of knowing that a thing full of life—a grasshopper that is able to leap into the air in one bound—can be reduced at one's whim to a bundle of quivering tissue.

Even children, then, have great power in their limited sphere, and they have been known to abuse it horrendously. As for adults, their behavior can descend into pits of abomination in totalitarian institutions such as slave-owning estates, prisons that mix habitual criminals with first-timers, the insane asylums of an earlier age, and, egregiously, the Nazi concentration camps of our time. In concentration camps a chasm yawns between the SS guards and prisoner-functionaries, who wield absolute power, and their victims, who are utterly helpless—just dumb animals and raw material to be used and ultimately destroyed. And how is this done? Both systematically, as in death factories, and on the spur of the moment in a mood of playfulness. Documented accounts of capricious evil exert a snakelike fascination: Splash prisoners with water in winter so that they freeze into solid blocks of ice and the corpses then have to be chipped or thawed out for disposal. Push a water hose into a prisoner's mouth, then turn on the faucet full blast so that the pressure explodes the victim's innards; onlookers find the explosion greatly entertaining. Pack prisoners into a closet so that they are wedged in tight, with absolutely no room to move, close the door, lock it, stuff paper into the keyhole, and then go for a cup of coffee—that is, forget about it, except for the awareness that the prisoners will suffocate and kill one another with body heat, in their desperate competition for air and in the crush of their writhing bodies.[11]

Cruelty in Primitive Societies

In "civilized" societies the same sort of disciplined imagination that creates wonders of intellectual and spiritual uplift is able to create unspeakable horrors of degradation—literal hells. In disgust, we look to earlier times and simpler (more primitive) societies for reassurance. Can it be found there?

"Primitive" is now considered a derogatory word to be avoided by speakers of good will. But its root meaning has no bad odor. Rather than its current sense of "crude," it meant simply "early" or "original." Thus, the primitive church is one where doctrine has been preserved in purity, uncontaminated by later vain excrescences. Primitive people are people who have retained the innocence and virtues of an earlier golden age. But if "primitive" as crude is not to be used, is "primitive" as pure apt? Are there—have there ever been—such unfallen people? Books that romanticize them regularly appear. They do not, however, tell the whole story. In such books dark facts, although they are known to the author, tend to be downplayed. What are these facts? What is the nature of the primitive people's offense? Their offense is the all too commonplace sort that the "normal" commit against the "deviant," the strong against the weak.

All societies contain the strong and the weak, and the weak are inevitably at the mercy of the strong, unless, *mirabile dictu*, a high collective standard of justice is conceived and imposed. Consider the Mbuti of the Congo rain forest. They have been idealized by

Colin Turnbull in his widely read book *The Forest People.* The Mbuti indeed have many endearing traits, outstandingly their warmth toward one another. But outsiders, such as the Bantu agriculturalists who share the forest with them, are beyond the pale, not fully human, beings against whom cheating and stealing are permissible. Mistreating weak members of their own group also appears to be morally unproblematic, in the order of things. The Mbuti esteem verbal cleverness and despise dumbness, which they associate with animals. A young man who happens to be a deaf mute is the camp clown and mercilessly teased for his stuttering speech—"animal noises," as his fellow campers call it. The teasing goes beyond just high spirits. On one occasion, the deaf-mute youth climbed a tree with a companion to collect honey. The task has its risks, which are nevertheless worth taking because honey is for the Mbuti the forest's single most desirable gift. Back on the ground, the youth rattled his rusted tin cup and with gestures and noises indicated that he wanted his share. He was roundly ignored.[12] Would an appeal to justice, or fairness, have made any sense? Fairness does make sense within a kin group, or among equals, for reciprocity is the key to survival. But it does not apply to those with a physical handicap, who have nothing to offer the common life other than as figures of fun.[13]

The closeness of primitive people to the world of living things, particularly animals, has often been noted by Western observers. To an unsophisticated reader of anthropological literature, totemic classification can make it seem that the tribes that abide by it confuse humans with animals, that they treat them with great consideration because they are, after all, kin. Among primitive folk, whether totemic or not, animals are indeed shown consideration and respect, but also neglect and cruelty—contradictory behavior that such people share with "enlightened" modern men and women. Consider, again, the Mbuti. Turnbull admitted to feeling dismay when he found that this gentle folk could derive pleasure from watching wounded animals suffer agonizing pain. The domesticated dogs that they keep for hunting might be kicked "mercilessly from the day they are born to the day they die."[14] Arctic hunters are reputed to propitiate the animals they must kill; they are said to avoid inflicting on their prey not only unnecessary pain but also humiliation. To what extent this is true and derives from a genuinely humane ethics, a worked-out morality of reciprocal consideration, is problematic. Understandably, native informants are eager to tell Western enquirers what they want to hear, especially when what they want to hear are the pieties of the tribe. During the rebellious 1960s and 1970s, college-educated people in the West, disgusted with their own world, sought—and naturally found—innocence or moral elevation elsewhere. "Là-bas on était bien!" they might have said, as did their intellectual forebears, the philosophers of the Enlightenment. It is hard for idealistic Westerners to accept the reality that few places are consistently Edenic, that few people (if any) are free of the mark of Cain, that folk of simple material culture are not necessarily wholesome in all respects, that they can be cruel, that the Inuits, despite their belief in nonviolence, may laugh at or beat mortally wounded animals, that they may even encourage children to torture small animals and birds to death, that the peaceable Machiguenga Indians of the Peruvian Amazon may treat their hunting dogs with abominable cruelty—rubbing hot chili peppers in their mouths, forcing them to swallow, causing them to howl and writhe in agony, all as a sort of entertainment.[15]

Violence: Exploding the Other

Violence admits of degrees. At one end, it inflicts mild incapacitation, at the other, total destruction. Why total destruction? In wars, whole cities may be razed to the ground, their populations dispersed or killed en masse. Violence of this order would seem to arise from the compulsion to have a clean slate, a whole new beginning, with the integrity of the new dependent on the total removal of the lingering powers of the old. Civilization has always treated nature thus; entire forests are erased to build cities and roads, the integrity of the latter depending on the nonexistence or total compliance of the former. In eating, as I have noted repeatedly, a human being consumes nature to maintain his bodily integrity; storing up oneself at this basic level of necessity depends on the evisceration (the "explosion") of the Other.

What if the Other is another human being? In a face-to-face encounter, another person has power over me in that I risk becoming a mere object in his perceptual field (the Sartrean anxiety). Although I in turn can reduce him to an object in *my* perceptual field, I retain an awareness that I cannot totally penetrate his subjectivity—his possession of an "inside" that, so long as it exists, gives him a measure of power. Now, this is admittedly a dire view of human relationships. Sartre has been criticized for putting it forth. Still, many—perhaps even most—human relationships contain a hostile element; suggestively, the words "hospitality" and "hostility" have the same root.[16] Conflicts may be out in the open and violent. The Other is not just another human being with whom I have to rub along, but a deadly enemy, a longtime oppressor, the personification of evil. My response to him may then be extreme. Such an individual must be not just killed but utterly destroyed, turned inside out, exposed, so that no secret source of power remains. During the Russian Civil War, an officer stomped his former master to death and explained to Isaac Babel why he had done it: "With shooting—I'll put it this way—with shooting you only get rid of a chap. . . . With shooting you'll never get at the soul, to where it is in a fellow and how it shows itself. But I don't spare myself, and I've more than once trampled an enemy for over an hour. You see, I want to know what life really is, what life's like down [that] way."[17]

The strong urge to penetrate, take over, subdue, and explode is almost exclusively male. It is bound up with the desire to *know* in both an intellectual and a sexual sense. A scientist is driven to penetrate his subject, break it down, expose its core, and, in a curious way, assume its "subjectivity." Nobel laureate George Wald's advice to his students was that if they really wanted to get to the heart of molecular behavior, they had to be able "to feel like a molecule."[18] A really good scientist can expropriate, so to speak, a molecule's "soul." While the procedure that leads to such total scientific understanding is admirable, it is also highly intrusive and indeed bears a distant, though still disturbing, similarity to another arena of human behavior. I have in mind the psychological drive and behavior of the New Warriors of the post-Vietnam era, as described by James William Gibson. In old movies, Gibson notes, gunfights end with the victim collapsing, his death indicated by a small red clot on his chest. In the new movies of the 1970s and 1980s, the victim disintegrates in an explosion of blood and phlegm. The warrior opens up his enemy so that nothing is hidden. His own body, sheathed in glistening skin, black leather, or metal, remains not only intact but uncontaminated by the effluvia of another's decomposition. (Think of the scientist's immaculate white coat.) Yet the warrior somehow feels that he has absorbed the enemy's life force, which he can now redirect toward a good cause.[19]

Humans As Animals and Pets: Sadomasochism

Exploding another, as envisaged in New Warrior movies and stories, is pornographic violence—pornographic because the violence is colored and inflamed by lust. More subtle is the sadistic/sexual pleasure that the strong take in torturing or humiliating the weak, without going to the extreme of killing them. For this "game" to be played, inequality has to exist, which is not a problem, for inequality exists in all societies, although no doubt it is greater in advanced, complex societies, where technical resources can combine with trained imagination to make both the abuse and the submission widespread and ingenious. It is also true that in some advanced societies, deservedly labeled civilized, steps have been taken to promote equality and compassion, working against ingrained feelings and biases—against a sadomasochistic propensity that may well be a signature of our species. This "unnatural" ideal of equality and compassion, hesitatingly approached but never fully realized, will be taken up later. Here I probe the dark side, which can at times cast such gloom that the good news is forgotten or remembered only as fairy tale.

Abuse of the weak is a deviant offshoot of the more general human domination of nature. "Nature" is whatever lacks, from the viewpoint of a powerful elite, their own dignity—their endowment of consciousness and will, their ability to create or acquire artifice. Judged by these criteria, animals clearly belong to nature. They may be figuratively elevated, treated as emblems of power and glory in sacred art, but in actuality they are almost always put in a subordinate or humiliating position, as beast of burden, producer of raw material (honey, silk), valuable property (nature's fanciful art), or pet. When humans are reduced to animal status, they themselves become a natural resource to be used and enjoyed by others. For the enjoyment to be intoxicating and erotically tinged, the social chasm has to be deep and broad. As people at the top arrogate quasi-divine status to themselves, they seem to feel a compelling need to see the people below them as mere "nature." Jean-Paul Sartre makes the suggestive comment that only the elite in an agricultural civilization possess true luxury, for luxury goods are essentially rare, *natural* products. In modern thought, workers are people who transform natural products into artifacts, and gain dignity in so doing. In an agricultural civilization, however, human labor itself, upon coming into contact with nature, is reduced to natural activity. "In the eyes of the rajah," says Sartre, "the pearlfisher did not differ much from the pig that nosed out truffles; the labor of the lacemaker never made of lace a human product; on the contrary, lace made a laceworm of the lacemaker."[20]

The economic exploitation of human beings is well known. Less well known is their exploitation as comely performing animals, subjected to the whim and patronizing-cum-erotic gaze or touch of the powerful. The ways and means by which the powerless are humiliated are numerous—some brutal, others subtle and diabolically clever. Placed side by side, they fill a long, infernal gallery. Even to step into the gallery is to risk moral tainting. Nevertheless we shall do so, not of course to indulge our sadomasochistic voyeurism, but in the interest of scholarship.

Early in the twentieth century, pedestrians passing over Hungerford Bridge tossed coins onto the mud banks of the Thames. What for? Well, for the pleasure of watching slum children—known as mudlarks—dive into the fetid paste as though mud were their natural habitat. A more benign variation of this game is for passengers on luxury liners to throw money into the clear and shallow waters near tropical islands and watch the natives, nearly naked and lithe like porpoises, plunge in for their paltry reward. Close to sadistic taunt is the following story as told by the eleven-year-old Theodore Roosevelt. In the course of a grand tour of Europe, the Roosevelt family came upon a group of Italian beggars. Young Teddy happily reported, "We tossed the cakes to them and fed them like chickens . . . and like chickens they ate it. Mr. Stevens [a traveling companion] kept guard with a whip with which he pretended to whip a small boy. We made them open their mouth and tossed cake into it. We made the crowds give us three cheers for U.S.A. before we gave them cakes."[21]

Anecdotal examples need to be placed in larger historical-social contexts. A curiosity of Renaissance and early modern Europe (1500–1700) was the popularity of dwarfs and midgets as pets.

They might be kept by the dozen in palatial households, dressed fancily, fed well, smothered with indelicate kisses, passed from lap to lap in amusement, or offered to a powerful patron as gift.[22] Making pets of deformed humans, though also known to the court of the Ptolemies and to decadent Rome, was rather an esoteric taste. Far more common were the ways that slaves in a variety of civilizations, in both ancient and modern times, were played with. Which was worse—to be a field slave or a house slave? In the antebellum American South, slaves might well prefer the hard labor of the field to softer conditions in the house, where they would be subjected to their owner's constant demands, shifting moods, and capricious "kindness." Servants too had their use as pet and decor. In a grand Chinese household, young ones were regarded as cute animals. As one lady says to another in an eighteenth-century novel, "You can talk to them and play with them if you feel like it, or if you don't, you can simply ignore them. It's the same when they are naughty. Just as, when your puppy-dog bites you or your kitten scratches you, you can either ignore it or have it punished, so with these girls."[23] In Victorian England, upper servants were a type of decor, selected for their stately height and the shape of their calves. A pair of handsome footmen went with a pair of handsome horses and a fine carriage.[24]

Acceptance of One's Place

What is it like for humans to be treated thus—reduced to nature, working animals, or playthings? How do they respond to such injury and humiliation? With despair, resentment, acceptance? The answer is, With all three in varying degree. Hardest for modern democrats to acknowledge is the third and by far most common response: acceptance. If only acceptance were a consequence of the recognition of an implacable force and sheer necessity! But that has not by any means always been so. If initially there was a recognition of force, eventually and all too smoothly the necessity of submitting to it came to be confused with the good. In a caste society such as India's, for example, those in the bottom layers could endure their station because they came to see it not so much as oppression as in the nature of things—reality—and hence moral and just.

Caste India is a somewhat extreme case. It is hard for a free citizen to understand why an untouchable might actually resist liberation, be resentful—even morally indignant—at being invited to move out of his "rightful place."[25] But even in a free society, being "in place"— though not, obviously, in the rigid form that it takes in a caste society—is important. Just about everyone needs the reassurance of having a place, knowing his or her place, being someone of importance within a social rank and yet, thanks to a sort of swooning incorporation into it, no one so special and different as to feel isolated and vulnerable. An individual's uniqueness is more often experienced negatively, as being out of step and a target of envy or contempt, than positively, as a source of individualizing and creative power. Alone, with nothing to do, one feels unreal—hollow at the core. Having a place and an occupation, no matter how lowly, fills the hollowness. Moralists say that we are enormously complex, multidimensional beings and that we resent being reduced to some simpler entity. While this sounds plausible, in actuality most of us need to be so reduced—to be known and to know ourselves as father, mother-in-law, teacher, janitor, or whatever. Indeed, each of these roles, though already a reduction, can still offer too much freedom, for it still leaves open the question (not completely settled by society's rules), What sort of father? What sort of teacher?

The ultimate reduction is to the status of dirt, both in society's and in the afflicted person's own eyes, as with India's untouchables. But a caste system that includes an outcaste of human pollutants (dirt) is exceptional. Much more common is the employment of all sorts of less formal and less institutionalized means to turn society's powerless into, not dirt, but something more like "slave and animal"—that is, to reduce them to biological nature. At that level of reduction, almost all human freedoms are stripped, all hollownesses plugged. Would this be a cause for despair? Not inevitably. Even in that condition, necessity can be transmuted into good. Indeed, it is not all that difficult, for as chattel and domestic animal the victim is freed from the need to provide for himself and free to live "in innocence"—that is, with burdensome chores rather than burdensome choice, removed from the stings of metaphysical anguish, including the anguish of mortality and eternal oblivion. To be at last at one with nature is to *be* unreflective nature, and that can be a great comfort, especially if one has a touch of masochism in one's psychological makeup—and who hasn't?

These examples appear to have come from some dark age. We may reassure ourselves by saying that yes, the caste system, patronage (a society of patrones and peons), and treating people openly as mere aesthetic objects and pets were once—and not so long ago—a blight on humankind. But we have left all that behind. The question is, Have we? Caste and feudalistic relationships, as institutions, no longer exist. But as practice they linger, albeit in milder forms. Consider paternalism. Imperialism in its idealistic heyday was openly paternalistic; the lesser breed were to be treated strictly but kindly—for the good of civilization and empire, but also for their own good. The lesser breed were "the little brown brothers." "Little," "brown," and the portrayal of them as seminaked transform native people into domesticated animals and pets. Not just European but Japanese imperialists saw themselves as superior beings of light color extending a helping hand to the small brown man of bare torso (in the Japanese case, usually a Southeast Asian).[26] In a liberal democracy, such an attitude and such behavior are unacceptable. Nevertheless, they are not entirely absent. They take the benign form of a compassionate and sensitive leader, a defender of justice, helping the less fortunate members of society, who are seen—and this is where the attitude and behavior assume a darker shade—as *his* people, put upon, struggling, but also ignorant, fundamentally incapable of protecting themselves or having long-range plans of their own. The put-upon people, on their part, play the game for all its political worth. But their acquiescence in the subordinate role is not just strategy. There is a sense that they accept—even feel proud of—their modest status, captured in such expressions as "ordinary people," "the little guys," and "just plain folks." Claiming to be plain and ordinary is reassuring, for if one doesn't aim high, one cannot fall far. At the same time, "plainness" and "ordinariness" are a tacit claim to modesty and other solid virtues.

Acceptance of one's place would seem to argue against the urge to escape that I see as fundamental to being human. The contradiction is more apparent than real. Consider again escape from nature. When it rains, we dash into a house—into a haven of our own making; the move is from "space" to "place," from uncertainty to the known. So much of culture, I have argued, is escape from the threats and uncertainties of nature. But the cultural world can itself be full of threats and uncertainty. In a dynamic society even the freedom and opportunities it offers can be burdensome. Escape, then, is "diving under the quilt"—holding on to, or retreating into, warm pockets of custom and habit.

Bystander and Voyeur

I have drawn attention to violence. Above a certain stimulus threshold, animals and humans lash out. Unique to the human animal, however, is the aesthetic appreciation of violence and destruction. Humans can even become connoisseurs of mayhem. "What joy it is when out at sea the stormwinds are lashing the waters, to gaze from the shore at the heavy stress some other man is enduring!"[27] So Lucretius famously observed. But there is no need to seek support from literary sources. History provides countless examples, especially in the area of crime and punishment. In Europe, the torturing and hanging of felons were carnivalesque occasions that consistently drew gawking multitudes. As though these gruesome events in real life were not enough, France's popular theater of the fourteenth, fifteenth, and sixteenth centuries offered simulations that lasted even longer.[28] Since the second half of the nineteenth century, death penalties in the Western world have been carried out in the privacy of prison compounds. Morbidly curious elements of the public, not by any means a negligible number, feel deprived. When in 1976 a convicted killer, Gary Gilmore, was about to be executed, television newsmen begged the Utah prison authorities to allow them to film the event. If they refused, said one Salt Lake City television station representative, desperate for a real-life macabre show, the station would consider "using paragliders, long lenses, helicopters—maybe even a dirigible."[29]

For this interest to occur, certain conditions must be met. First, one must feel secure oneself. Safe on the beach, I can watch sailors drown in the raging sea with tingling excitement. Second, one must be able to identify with the victim. The greater the identification, the greater the interest. The chopping down of a tree is not so exciting to watch as the butchering of a pig, and still more exciting—a sort of gripping horror from which one cannot quite avert one's eyes—is the punishment, or in the extreme case the killing, of another human being. Moreover, one must believe that the fates, having already found a victim, will somehow leave oneself alone. And one must be in a state of mind that can, contradictorily, both identify with and dissociate from what happens "out there." Spectatorship is the name of the game. In the theater, a spectator, even when deeply engaged, does not have to—indeed, must not—act. In real life, the same situation may demand action; simply looking can be profoundly immoral.[30]

Displace and Disconnect!

Imagination empowers one to escape. How does *evil* imagination effect escape? One way is by displacing whatever is bad from self to others. Spectatorship is such a displacement. Seeing others in a shipwreck or car accident says to me that fate has struck elsewhere. I am here, not elsewhere; as a spectator I will always be safe.

Treating other people as animals is another such displacement. "Bad" in animal life is having to struggle for food, having sexual urges that detract from one's dignity, and having to die. Death is the ultimate bad—bad not so much in the endpoint itself as in what precedes it (suffering, pain, and the humiliation of being reduced to a body) and in what it implies (a human being's utter insignificance). To escape this, to elevate oneself above the neediness, frailty, and insignificance of mortals, one can wrap oneself in the trappings of divinity. These trappings may be material—a big house and other forms of wealth that distance their owner from nature. To really feel godlike, however, more is called for than material possession. That "more" is the actual exercising of power—if possible, absolute power. Absolute power creates the illusion of a radical gap between myself and others. To the one side is me, semidivine in potency; to the other side are human animals on whom my potency feeds. My own animality is forgotten, displaced from my core nature to the core nature of others. How can I die when I have power over other people's life and death? How can I be victimized by pain and suffering when I, with just a nod of the head, can cause pain and suffering in others?

Evil imagination seeks to displace. More widely and diligently, it seeks to disconnect. It draws boundaries, an important purpose of which is to shield me from contamination by other people's misery and misfortune. The boundary must be all the more sharp and firm if I myself am the cause of that misery and misfortune. A place seamed with such boundaries is hell. On earth, a pure expression of hell is the Nazi concentration camp. Most places are obviously not hell; most human lives are reasonably moral, pleasant, and open. Nevertheless, boundaries, though not impenetrable ones, are a commonplace of existence; they have to be if human lives are to be bearable and to move forward in a confident and orderly manner. The ability to disconnect, to separate the self from others, to separate even the different roles and faces of a single self, may be morally problematic, but it is also a condition of sanity.[31]

Consider what happens in the course of an ordinary day. I undergo a succession of changes, the nature of which depends on where I am and the people I am with. At the farm I fondle a lamb, muttering sweet nothings into its ear; in my own dining room only a few hours later I chomp appreciatively on lamb chops. One moment I am in the men's room, my face red and sweaty with the effort to evacuate; the next I am standing in the classroom, coolly lecturing on Platonic ideals. Daily life is full of such shifting scenes, and it would not do for me to carry the self of one scene into the next. Mostly I don't. I am able to skip across boundaries into new roles because, like other functioning individuals, I can disconnect—forget—cleanly.

Society helps by telling people when and where to do this or that, with boundaries of varying firmness drawn between the whens and wheres. All societies? I would say yes, for all societies have rules and taboos that require the segmentation of space and time. Such rules are taught to the young; their use in avoiding social conflict seems reasonably clear. Rules and taboos for avoiding moral and conceptual dilemmas are more obscure in ultimate purpose. An old moral dilemma is the killing of an animal that has been a human benefactor. Taboo resolves the dilemma by dictating that bodily desires and the moral sense be kept separate. For example, among cattle-raising Africans, eating meat and drinking milk at the same time is taboo. The proscription helps them to circumvent the untenable moral position of killing and eating the meat of an animal that has nourished them with milk.[32] But is this the solution of a problem or just the hiding of it—escapism in a reprehensible sense?

Whereas disconnecting one type of thought and activity from another is human, the extent and rigor with which it is carried out vary. Generally, the more developed the society, the greater the felt need to disconnect, beginning with the construction of walls and walled rooms in Neolithic villages and progressing to the increasing partitioning of space—the construction of all kinds of rooms that cater to all kinds of needs and activities—in great cities. People learn to change behavior as they move from one room to another. They may be said to become different selves in different rooms; what is socially or morally proper in one may be socially and morally improper in another. But this shift of attitude and value is forgotten; change in setting makes it easy to forget.[33]

Is *dis*connecting a necessary condition to making proper and energetic connections later? As a social and moral being, I have to connect. A father, I feel my son's frustration over his roller-coaster romance; a teacher, I sympathize with a colleague's heavy schedule of work. These are more than just feelings, for I may spend time and energy to help. On the other hand, to have such resources on call I must know how to conserve and direct them. If I seriously tried to help the homeless on my way to school, I would be too exhausted to be of use to students. So I toss small change and pass on. In public spaces I look at people, if at all, blankly; later, in my own classroom, I am attentive and caring. The way I report these behaviors makes them sound reprehensible. Maybe they are, judged by the highest moral standard. Such a standard, however, is an abstraction quite dissociated from how people actually behave, or even from how they believe they ought to behave. In premodern villages and towns, turning a blind eye on needful strangers is quite acceptable, even commendable. A man's obligation is to his family first, neighbors next, and strangers last; altering the priorities would be unnatural. Such was the Chinese view under Confucian influence. Confucians deserve credit for at least saying that strangers should have the leftovers.[34] Elsewhere—in African and American Indian societies, for example—even that minimal obligation is not recognized. A rich tribe—rich, say, from oil and coal revenues or the proceeds of casino gambling—feels no obligation to extend financial aid to a poor tribe, even though both may unite in denouncing the lack of generosity on the part of the federal government.[35] Restricting charity to one's own community is adaptive. Not recognizing, not sympathizing with, the sufferings of outsiders makes good sense in times of scarcity; and historically, times of scarcity have been all too frequent, even in well-favored parts of the world.

Dissociative Monstrosities

These examples of dissociation—escapes from the burden and embarrassment of connectedness—are from ordinary life. They do not, as a rule, trigger moral alarm. Yet out of habituation to them dissociative monstrosities can emerge. One all too familiar type is the wall that separates ideal from real, vaunted belief from practice. All societies that have embraced some high ideal suffer in some degree from such disjunctions—egregiously, Europe in the fourteenth century. It was then that "knights talked about chivalry, but regularly practiced treachery, murder, rape, and rapine. The pope walked up and down reading his breviary to the background accompaniment of the screams of cardinals being tortured to extract confessions about a suspected plot against him. The Church preached poverty and chastity, but the high clergy openly wallowed in opulent luxury and endowed their bastards with the spoils of fat clerical livings, while the poor friars enjoyed an apparently deserved reputation as habitual seducers of married women."[36]

Behaviors and actions too can be rigidly compartmentalized, as though they were different realities altogether, with no bearing on one another. I have already noted how people switch roles easily, act at different times as though they were different persons. Perhaps this is because, at a deeper level, the human psyche is not a whole but rather, as Walt Whitman boasted, "a multitude."[37] "Multitude," while it pleasingly suggests richness and flexibility, disturbingly implies the absence of a unified personality. What we see as an individual person may well be a promiscuous mix of selves—liar on one occasion, truth teller on another, hero in the morning and coward at night, lecher and puritan, sinner and saint, in quick succession. An extreme example will demonstrate that these changes of outlook and role cannot be dismissed as socially necessary, or as merely amounting to colorful inconsistency and playful, occasionally irresponsible, behavior. My extreme example—and probably everyone else's—is the rigid moral compartmentalization as practiced by SS guards in concentration camps. Even against a background awareness of our own shifts of role and personality, it is still a shock to know that the guards who nursed their sick dog in the morning and looked forward to playing a Mozart quartet in the evening betweentimes had men, women, and children shoved into the gas oven.[38]

Consistency and connectedness are intellectual virtues. One might therefore think that they are more often to be met with in academia than elsewhere. Not necessarily. A university is a historical institution made up of fiefdoms (disciplines), the older among which, deeply rooted in tradition, are jealous of their independence. Moreover, within each fiefdom there are enclaves (subdisciplines) that make little effort to cross their own self-defined boundaries. This is one reason—the historical reason—for the lack of connectedness in universities, for the disinclination to seek inspiration elsewhere. But there is another reason, more psychological than historical, which suggests that human beings have a limited amount of moral/intellectual capital. Spending it lavishly in one area can leave other areas famished. In the university, specialization, which is the main road to progress, invites that imbalance to a high degree—at least, according to Robert Hutchins, one-time president of the University of Chicago. He wrote bitterly, "The narrower the field in which a man must tell the truth, the wider is the area in which he is free to lie. This is one of the advantages of specialization. C. P. Snow was right about the morality of the man of science within his profession. This is because a scientist would be a fool to commit a scientific fraud when he can commit fraud every day on his wife, his associates, the president of the university, and the grocer."[39] Hutchins surely exaggerated. Spending eight hours every day evaluating evidence that can dash one's hope for a desirable outcome does tend to make one more circumspect and honest in adjoining areas of life. But Hutchins has a point; the contrary can and does happen, more often than we care to admit.

An Isolated Good Action: What Is It Worth?

A scientist may be selfless and truthful, but only (so to speak) by the hour. Such hours may be enough to make him a good scientist but not a good human being, for a test of goodness in a human being is consistency. In *The Brothers Karamazov*, Grushenka tells the fable of the little onion. A vicious old woman dies and goes to hell, but her guardian angel, straining his memory, recalls that she once, only once, gave a beggar the gift of a little onion she had dug up from her garden. He holds the little onion out to her, and the old woman grasps it and is lifted out of the flames of hell. Primo Levi, who had known the horrors of the Nazi concentration camps, was outraged by the fable. He called it "revolting." What human monster, he asked, did not sometime in his life "make the gift of the little onion, if not to others, to his children, his wife, his dog?"[40]

An isolated gesture of good will can be capricious, or, to put it another way, as easy and automatic as a good sneeze. Alone it says nothing of a person's moral health. Being good in the moral sense involves commitments and liabilities that reach far beyond the odd kind act. In this regard a good act differs importantly from a good work of art. A work of art stands by itself, beautiful no matter how many ugly pieces precede or follow it. The artist, despite many mediocre productions, is still an artist, even a good one, if he or she has one outstanding success. Failed works do not mean a lack of good intention; moreover, they may help the artist create a masterpiece in the end. Evil deeds, by contrast, are driven by mixed if not downright evil intentions, and are in no sense a preparation for the rare good deed. The aesthetic life is one of discrete vivid moments—a beautiful sunset, a lovely symphony—separated by long stretches of gray. In contrast, the moral life is one of maintaining a continuous charitable relationship—sometimes happy, often frustrating and sad (the sadness of inevitable misunderstanding)—with others. And this in turn suggests that a good person cannot be a "multitude," a medley of roles. He or she has the sort of dependability that being a unified whole, a wholesome (holy) person, implies. No wonder novelists find it hard to make a good character interesting. No wonder, in fictional works, the average sensual guy and, *a fortiori*, the villain are the heroes.

Idols

Compartmentalization is one way to escape the clash of conflicting desires and ideals. Having idols and worshiping them are another. Indeed, the two are similar. An idol may be regarded as a sort of compartment into which an individual or a whole society puts its emotional treasures, making it possible to deposit rival goods in side drawers where they are conveniently forgotten. The idol may also be seen as an anchor, a focus, or a fount. Whatever figure is used, it provides reassurance and comfort in a world that, even when it is generally well run, can still seem at times bewildering and frightening. Material plenty eases some of that bewilderment and fear, but not all. Reflective individuals know that their sense of well-being depends on other, less tangible things, including—and this can be deeply troubling—an ability to maintain, unreflectively, the partitions of life and thought.

The need for idols is a sign of imaginative exhaustion. A golden calf is so much more manageable than God. A human idol makes even fewer demands on the imagination, for he or she can issue orders, promulgate rules of behavior, insist on adulation, whereas a golden calf does so only by proxy or through the idolater's own overheated imagination. Human beings—Napoleon, Elvis Presley—are among the most popular idols. Wealth too, for it requires little imagination to see it as the fount of well-being. Wealth takes different forms. A house represents wealth, as does also, more abstractly, a bank account. An idolater wants both to grow. But even within the world of wealth, different forms call for different levels of imagination. Concrete things, surprisingly, can make a greater demand. A bigger house is not necessarily a better house, whereas a bigger bank account is self-evidently better than a smaller one. Not all idols are material. Prestige, for example, is an idol only loosely tied to its material embodiment, which may be as fleeting and intangible as a gesture, a tone of voice, a fragrance, and as easily missed as the tiny insignia on a person's shirt. The acolyte of prestige needs to be constantly alert and of subtle mind.

Religion, properly understood, is the opposite of all that the word "idol" implies. Yet it is more cluttered with idols than any other sphere of life. The golden calf conveniently stands for all the myriad objects, from Buddha's tooth and the true cross to great works of art and architecture, before which humans are inclined to worship, to treat not as emblems of the divine or windows to transcendence but as ends in themselves or powers in their own right. Material objects are one part of the paraphernalia of religion. As important are the gestures, rules, and taboos. Circumambulating a shrine the wrong way, eating the wrong sort of food, can seem the utmost impiety. It is as though the majesty of God, the salvation of souls, and the harmony of the universe all hinged on and could be reduced to correct performance. Idol worship, as distinct from true religion, is nearly irresistible. We can see why. It is easy to do, and this despite—and even on account of—the rigid rules that have to be followed. It is also easy to do because its sacred objects are tangible, as God in his otherness—"not this, not that"—is not. Furthermore, obeying injunctions to the letter provides two great psychological rewards: self-righteousness and security. Obedience in itself can generate a sense of self-righteousness; as for security, the rules that confine one are like the sheltering walls of a house.

Madness or Exhaustion

Compartmentalization and disconnectedness appear to be intellectual and moral defects that need to be overcome. "Only connect," E. M. Forster famously urged.[41] And one can see his point. So much of human ill lies in our inability to connect—to empathize and sympathize, to stand in another's shoes and see another's viewpoint. On the other hand, how far can feeble human beings go in that admirable direction? Moralists may command us to be "involved in mankinde," but that path, if conscientiously followed, leads straight to madness. In fact, one understanding of schizophrenia is that its sufferer lacks protective walls, is too much inclined to feel with others. This inclination is involuntary—the taking on of heroic sainthood in spite of oneself.[42] Sanity requires firm boundaries that monitor the assault of stimuli, many of which are unpleasant and some of which are horrendous. Not only that. One's usefulness to another depends on a certain coolness and distance. Obviously, no surgeon who feels the pain of the patient can successfully operate on him.

Less emotionally charged than empathy is sympathy. We ought to open ourselves to another's likes and dislikes. Educators constantly urge us to savor the strange and the challenging. But here too we are constrained by our weakness. In an eloquent speech decades ago, the brilliant physicist J. Robert Oppenheimer confessed to his own sense of exhaustion in the "great, open, windy world," one in which "each of us, knowing his limitations, knowing the evils of superficiality and the terrors of fatigue, will have to cling to what is close to him, to what he knows, to what he can do, to his friends and his traditions and his love, lest he be dissolved in a universal confusion and know nothing and love nothing." We may have to leave the room when someone "tells us that he sees differently from us and finds beautiful what we find ugly." But that is our weakness and our default. "Let it be a measure of our virtue that we know this and seek no comfort."[43]

Dousing the Mind: Escape to the Bottom

The more people use their mind and freedom, the more they may be tempted by the bliss of nonexistence; the higher they fly, the more appealing can seem the peace and stability of the hole or bottom; the greater their power, the more they may yearn, if only temporarily, for powerlessness—total submission to the will of another.

John Keats is the poet of intensity, both sensual and intellectual. Not surprisingly, he is also the supreme poet of lethargy. How can any harassed and painfully alive person read the following lines and not feel their gentle pull?

> Ripe was the drowsy hour;
> The blissful cloud of summer-indolence
> Benumb'd my eyes; my pulse grew less and less;

Pain had no sting, and pleasure's wreath no flower.
O, why did ye not melt, and leave my sense
Undaunted quite of all but—nothingness?

The poet confesses to being "half in love with easeful Death," to calling him "soft names" and longing "to cease upon the midnight with no pain." The eloquence is exceptional, but who, after all, has not known the temptation? Even William James's healthy-minded man cannot be wholly a stranger to Hamlet's wish "to die, to sleep . . . to die, to cease."[44] Those less robust live with it as a promise and a friend. In a short essay, "Sleep, Sweet Sleep," the thirty-three-year-old Thomas Mann claimed to remember every bed he had ever slept in, that he loved sleep even when he had nothing to forget, that he yearned for the dissolution of tyrannical will. Sleep was erotic bliss, its foretaste indolence, its climax death. Not surprisingly, Mann admired *Tristan und Isolde* above all works of art because of its "yearning after holy night."[45]

Death is final. Of the less extreme ways to douse the mind, drugs, including alcohol, are readily available and commonly used, but even closer at hand is the imagination, which amply provides its own soporific. So it's tough being a human; well, what about turning oneself into an animal that chews the cud like a contented cow or is able to withdraw into its shell like a turtle? "I believe I could turn and live with animals," Walt Whitman wrote, "they are so placid and self-contain'd."[46] Animals don't have to meet the moral demands of their fellows. Humans do, and even more burdensome for them are the relentless demands they place on themselves. Reversion to less sentient forms of life has a persistent appeal, as myths show. A major early source is Ovid's *Metamorphoses.* It is rich in stories of perplexed and wounded people who yearn to be turned into animals or even into plants. Fascination with children brought up in the wild by animals is part of the same wish, only less extreme. The question here is, What is it like to have human faculties and yet live without the benefits and burdens of human culture? When a "wild boy" was discovered in southern France at the end of the eighteenth century, he inspired scores of articles in which his contemporaries spoke of "waking this creature from his sleep." "And so they did up to a point," wrote Roger Shattuck. "Yet one also senses that many people were drawn to his unthinking existence as if they, too, yearned to lay aside their everyday identities and responsibilities."[47] Perhaps the mildest and most commonly used soporific among romantically inclined intellectuals is to imagine the jettisoning of sophistication in favor of the life of a peasant. An outstanding modern example is Tolstoy, who in his ceaseless copulations with peasant women in his youth, and in his mature idealizations of the simple man of the soil, sought to escape the monitory awareness that tortured him throughout his life.[48]

Peasants are poor. Poverty may be a cure for metaphysical angst. But whereas the middle class are free to fantasize about the wholesome life of peasants, they dread the deprivations of poverty. George Orwell tried to reassure them. Orwell was a member of the "lower upper middle class" (his own categorization) who for a time lived as a "down-and-out" in Paris and London. The tumble to the abyss was horrible, but, to Orwell's surprise, once he hit bottom and settled down, life was unexpectedly tolerable. For this to be true, however, two conditions have to be met: one must have no family responsibilities, and one has to have hit rock bottom. As Orwell explained it, a man with a hundred francs may go into a state of craven panic, but with only three francs he says to himself, "Well, they will feed me till tomorrow," and he can't think further than that. Poverty annihilates the imagined worries of the future. That is one source of relief. Another, which is almost pleasure, is "knowing yourself at last genuinely down and out. You have talked so often of going to the dogs—and well, here are the dogs, and you have reached them, and you can stand it. It takes off a lot of anxiety.” In poverty, one becomes an urban forager. That endless search for food and shelter certainly complicates life, yet it also simplifies life by channeling the mind to just these urgent and immediate goals.”[illegible]

Masochism too is a way of hitting bottom, of escaping—a far more inviting way than actual poverty, for masochism is fantasy. “It is amazing how often rich and spoilt businessmen fall victim to masochistic impulses,” wrote Ferdynand Zweig, a sociologist. “This comes out very clearly in the stories one hears from prostitutes about the requirements of well-to-do clients. . . . Masochistic practices, they report, are especially prevalent among successful businessmen, and much more frequent than sadistic practices. Many well-off clients enjoy being beaten up, kicked, or whipped.”[50] Masochism allows one to have the best of both worlds: the world of power in normal life and the world of total submission in enacted dream life, which nevertheless is an indirect assertion of power, for the groveling scenarios are designed by the masochist himself. Masochist psychology may be bizarre, but it is not beyond the range of ordinary human desire. And what is that? It is the desire to be esteemed and yet able to savor, from time to time, the pleasures of abasement.

Fantasies of Flying

Bondage to earth provides security. At the other extreme is the security of total freedom. And what greater sense of freedom is there than flying? Who hasn’t dreamed of it—of being a bird, a skylark, as Shelley imagined?

> Higher still and higher
> From the earth thou springest,
> Like a cloud of fire;
> The blue deep thou wingest,
> And singing still dost soar, and soaring ever singest.[51]

Flying is a common, perhaps universal human wish. It appears in children's daydreams and in adults' myths and practices—in healing trances and ecstasies, shamanistic voyages, Icarus's melted wings, Leonardo's drawings of flying machines, and legions of angels and winged creatures.[52] With the invention of machines that actually do fly, dreams of defying gravity have not come to a stop; rather, they become wilder with each success, skipping from one achievement or possibility to another—from just taking to the air to swirling and diving like a bird, from leaping across the English Channel to leaping over continents and oceans, from reaching the moon and (imaginatively) planets to reaching stars and galaxies.[53] Was there a steady gain in the experience of freedom as ever more powerful machines were built? It would seem not. Modern airplanes fly high and fast, but to passengers strapped into their cushy seats they have the feel of lumbering crates. A primitive airplane, by contrast, offers greater adventure and freedom, its thin metal casement being less a barrier than a means to feeling the rush of wind and stars.[54] At the other extreme of technological ingenuity is the spacecraft that can sail beyond the orbit of the moon. No human has traveled in one. Nevertheless, it contributes enormously to our sense of freedom. Dull indeed is the man or woman unstirred by the adventures of *Pioneer 10*. Launched on March 3, 1972, it was by 1996 more than 2.5 billion miles beyond the edge of the solar system, which means that when the sun becomes a red giant, expands and gobbles up all the planets, there will still be this frail human contraption on its way to the next star, a tiny material witness to immortality.[55]

Isn't freedom from bondage to earth—a home that is also a tomb—a desire for immortality? The real gain with ever faster airplanes is the compression of distance, which ought to mean—but doesn't—an expansion of time. Time feels more than ever scarce. Yes, we fly, we aim at the empyrean, but in the end we yield to gravity, come down from our perch, which is never more than temporary, and die. That's the rub.

Fantasies of the Good

Are there other escape routes? Yes, surely many others. I will present one that is popular in modern times and that is strangely ambivalent, arousing both strong approval and disapproval. It is this: Rather than descend into dimmed states of consciousness or simply taking flight, imagination constructs glittering fantasies of the good. Such fantasy worlds may be just private daydreaming and go no further. They can also be embodied in writing, for example, fairy tales. And they can be embodied in architecture, from pristine suburbs to magic kingdoms inspired by Walt Disney. In any sober estimation, these are far from hell or hellish. They are indeed a type of the Good. Yet critics have attacked them as unreal or sinisterly real—hyperreal.[56]

Fairy tales are flights into fantasy, archetypal escapism, of use only as distractions and entertainments for young children. Adults tend to take this view for granted, even though it does not do justice to the best works written in the nineteenth and twentieth centuries. Of course, all fairy tales are fantasies; they are alternative worlds, not pictures of the familiar one. And they do offer temporary escape from the dreary reality of daily living. Why not? Escape from one kind of reality may be the only means of making contact with reality of another kind. For instance, withdrawing into the ivory tower—a university, a research laboratory—may be the only way to engage with difficult truths concerning the nature of the universe. The most unfair criticism of fairy tales is that they are wish fulfillment—a bad, escapist habit. Wish fulfillment is, however, not at all characteristic of modern fairy tales. The criticism may be better leveled at their eighteenth-century precursors. These indeed contain elements of unabashed wish fulfillment and escapism. Escape from what? Invariably it is escape from necessity, which at its most urgent is the struggle for food. Typically, once the peasant-hero gets hold of a magic wand or ring, his first thought is of food, and, if he is ambitious, not just any food but meat.

Nevertheless, old fairy tales are harshly realistic in one area: human relations. To survive, a father may sell his daughter because she eats too much and does not work enough. A hero bent on escaping with the princess thinks nothing of refusing help to a drowning beggar. Tales of this type are not so much immoral as amoral. Cruelty is taken for granted when the social order itself is cruel; cunning rather than courage or wisdom is the esteemed virtue. Another hallmark of such stories, as distinct from modern fairy tales, is a Rabelaisian delight in the functions of the lower body. The story "La Poupée" characteristically combines scatology with a lesson in mistrust. A simple-minded orphan has a magic doll that excretes gold whenever she says, "Crap, crap, my little rag doll." She buys chickens and a cow and invites neighbors in to see what she has. A greedy neighbor steals the doll, but when he says the magic words, the doll craps real crap all over him. "So he throws it on the dung heap. Then one day when he is doing some crapping of his own, it reaches up and bites him. He cannot pry it loose from his *derrière* until the girl arrives, reclaims her property, and lives mistrustfully ever after."[57]

One can imagine the scene in a village square as the story unfolds, the teller's lively speech made livelier by lewd gestures, his audience showing its appreciation with loud guffaws and backslapping. In our time the old fairy tale's piquant mix of extravagant fantasy and reductionist realism—humans reduced to the status of cunning animals—is much admired among literary pundits influenced by a Marxian distaste for bourgeois fantasy. And what is the essence of that fantasy? It is a world without crap, where "crap" stands for animality, earthiness, bondage to biological exigencies, death. Unlike the common people's dream, the bourgeois dream is expurgated; all that remain are material plenty in a lovely setting, good fellowship, and dignity for all but especially for oneself—that central figure around which all the delightful views, goods, and services seem arranged.

Such, however, is the power of the bourgeoisie that its dream can be (and has been) turned into reality; and so the more affluent parts of the world now boast ranchhouse suburbs, mega-shopping malls, and theme parks filled with carefully designed and marketed comforts, wonders, and thrills.[58] They are undoubtedly popular. Yet a small minority—usually those exceptionally privileged in education—do not want them. To this small minority the giddylands of modern consumerism are escapist fantasies because they deny the forces, which can be brutal, that have made them possible and also because they deny people's animal nature. Even if these places can be built without undue exploitation of either laborers or natural resources, they are unworthy—surface-deep and tendentious—unless they make allowance for filth. Of course, no one actually wants to be bogged down in filth, just to retain an awareness that it is irreducibly there. Even young radicals who disdain the picture-perfect suburb or mall cry "Shit!" when things miscarry, as though they had had enough of messy, unpredictable reality and would not mind being in a house where the coffee maker actually worked and the bedsheets smelled nice. I am led to conclude that the one element that is critically lacking in these places is a certain moral and intellectual seriousness. But isn't this just a personal or cultural bias? Is it realistic, or compatible, to call for moral and intellectual seriousness on top of the comforts and entertainments that such good middle-class places already provide? Straining after an image of the ideal place—utopia or heaven—is undone by two glaring contradictions: the necessity for crap (else how can any place be real?) and the necessity for some sort of continuous spiritual/intellectual aspiration and development that is at odds with who we humans *really* are.

5

HEAVEN

The Real and the Good

• • •

Practical life, repeatedly tested at the bar of survival, is indubitably real. As for conjurations of the mind—demons and ghosts, fairies and angels, gods and goddesses—we outgrow them as we gain more confidence in the material conditions of life and as we become better acquainted with the inner workings of the mind. Supposing this is true, at least for mature and rational adults, then how should such adults—how should we—regard myths and rituals, art and artworks? Are these things the epiphenomena of buried neural processes, pictures of the real, comments on it, embroideries of it? Or would it be better, less committal, to describe them as separate objects of human creation added to the totality of an ongoing and evolving reality? Furthermore, what are we to make of mankind's proudest achievements—moral ideals, elevating philosophies, great religious insights and systems of thought? Are they the ultimate fantasy, or are they, to the contrary, the ultimately real?

In the fifteenth century, Nicholas of Cusa, while discoursing on the fundamentals of science, could still admit magic and the kinship of humans with spirits. Francis Bacon, a century and a half later, went notably further in the direction of modern scientific thought. He rejected magic, equating it with dreams, hallucinations, and fantasy. He also rejected astrology and alchemy, for to him these depended too heavily on imagination and faith and were too remote from the experiences of everyday life. Bacon did, however, retain a theory of angels and spirits among the fundamentals of his "primary philosophy." Notable in this philosophy—an early signal of the Enlightenment—is his turn toward a brighter world. The supernatural residuals of his thought consist of angels rather than demons, spirits rather than ghouls.

Bacon's shift might prompt one to ask, Why retain angels? Are they any more real than demons? Most people probably consider the embodiments of evil more firmly grounded in experience, for life is undeniably harsh, often cruel. Even in the world of sleep, the monsters that haunt a nightmare are far more intrusively "real" than the good spirits that may appear in a pleasant dream. On the other hand, could Bacon be right? Although it may still be healthy realism to give greater credence to "evil and hell" (so much a part of life) than to "goodness and heaven" (so subject to wishful thinking), is it just possible that angels are more real, or have become more real, than demons? For it is a fact that monsters and ghouls, devils and witches, have faded in modern times. By contrast, God and spiritual beings have not altogether lost their respectability in serious thought, at least until late in the twentieth century. What it means to be a truly good person—someone both in this world yet otherworldly, almost like a visitor from beyond—is probed in modern Western literature; and God is still a part of the curriculum in theological schools. Imagination, after all, is not only a source of illusion and error, it is also the uniquely human path to knowledge. If in modern times one kind of knowledge—scientific—enjoys exceptional prestige, a reason more basic than usefulness is that it appears to present the truest picture yet of reality, as distinct from some fear-driven or wishful image of it. Surprisingly, this picture is not pedestrian, close to what common sense would offer, but full of wonder. The strangeness and wonder at the core of nature inspire even agnostic scientists such as Albert Einstein and Stephen Hawking to use the word "God" in their popular works and personal reflections; and this God that they proclaim has far more in common with a creator who impartially makes the sun "shine on the just and unjust"—sublime and ineffable—than with the morally ambivalent, impassioned, partisan deities of countless myths and folk tales.[1]

Fantasy and Science

For reasons still not well understood, sometime in Europe's early modern period a number of adults, like uninhibited children, started to ask questions that challenged well-established facts. The result is the formulation of some of the fundamental laws of modern science, outstandingly (to use a schoolbook example) the laws of motion—Galileo's and Newton's. Against common sense and the authority of Aristotle, Galileo discounted the effect of friction, though anyone can see that it operates everywhere and must affect the speed of a moving body. In a frictionless world—but what fantasy is this?—it can no longer be a fundamental principle that bodies fall with velocities proportional to their weights. Newton's first law states that a body in motion tends to remain in motion at a constant speed in a straight line unless acted on by an outside force. To the layman, that idea can still seem a willful fantasy postulated by a science-fictional writer determined to shock. For so far as observation goes, every trajectory—from that of a tossed stone to that of migratory birds to that of the sun—is curved or circular. Historically, the circle has had an extraordinary hold on the human imagination—ancient Greek imagination, certainly, and that of their intellectual descendants, the Europeans, right up to and including Copernicus and Galileo.[2] In other cultures and civilizations as well, not least the American Indian nations, one can discern the circle's grip. As Black Elk, an

Oglala Sioux, firmly puts it, the "power of the World always works in circles, and everything tries to be round."[3] Maybe he is right in some deep and mystical sense. Yet it is Newton's "unnatural" postulate of the straight trajectory that is to pass the one supreme test of the real, namely, power. And what does power mean here? It means, on the one hand, being able to identify connections in phenomena formerly taken to be different and discrete, and, on the other hand, being able to predict their behavior accurately with hitherto unknown economy.[4]

A more surprising case of how fantasy can take one to a vast new continent of the real is given by the historian of science Lynn White Jr. He writes:

> In 1733, as intellectual sport the Jesuit Girolamo Sacchari challenged Euclid's axiom of parallels, and substituting the "nonsensical" axiom that through a given point two lines may be drawn parallel to a given line, he constructed a self-consistent non-Euclidean geometry. Unfortunately, he laughed it off as a *jeu d'esprit*. Not until four generations later did mathematicians realize that he had made a great discovery. Then a whole constellation of contrasting geometries burst forth, and it was with the light of Riemannian geometry that Einstein found the mathematical key for the release of atomic energy. The most astonishing part of the new canon of symbols is the discovery that we human beings can deal with facts only in terms of fantasies.[5]

The idea that there is a "mathematical key" to nature was scarcely known before the sixteenth century. Only in the years immediately preceding 1600 was the mathematical equation itself being developed. Yet once it came into use, it was quickly acknowledged to be not just a powerful exploratory tool but the test of true understanding. The idea emerged that the fundamental nature of the universe is mathematical, that indeed God is a mathematician and that His works, if searched deep enough, reveal a precision and elegance uniquely captured by that austere language. How can it be otherwise as humanity outgrows its childish visions of God? We are still in the grip of this myth and may smile appreciatively at the story of how the schoolboy and future computer scientist Alan Turing grew immensely excited when he learned of the appearance in nature (for instance, in the leaf arrangement and flower patterns of many common plants) of "the Fibonacci numbers—the series beginning 1, 1, 2, 3, 5, 8, 13, 21, 34, 53, 89 . . . in which each term was the sum of the previous two."[6]

Biological nature that is accessible to the naked eye is actually a tangled mess, and whatever mathematical order there is may be just coincidence. Physical nature—the stars above—is by comparison more orderly, but even there modern scientists tend to discover chaos, historical accident, and violence rather than the simple harmony taken for granted by their predecessors.[7] Still, physicists have not given up. The test of profound truth is still beauty. As the astrophysicist S. Chandrasekhar puts it, "In my entire scientific life, extending over forty-five years, the most shattering experience has been the realization that an exact solution of Einstein's equations of general relativity . . . provides the absolutely exact representation of untold numbers of massive black holes that populate the universe. This 'shuddering before the beautiful,' this incredible fact that a discovery motivated by a search after the beautiful in mathematics should find its exact replica in Nature, persuades me to say that beauty is that to which the human mind responds at its deepest and most profound."[8]

In the world-view I have just sketched, beauty is explicitly, good is implicitly, recognized. Good, truth, beauty, and, yes, God are so linked that one cannot engage fully in the one without bringing in the others. Perhaps this way of looking at reality is just an intellectual habit that the West has inherited from Plato. Still, its perdurance is remarkable—and most remarkable of all is the support it apparently receives from modern science. Whatever standing the grand Platonic forms still enjoy today is owed more to scientists of a philosophical bent than to professional philosophers.[9] A professional philosopher brought up on late-twentieth-century relativistic and deconstructionist thought would be embarrassed by the label "seeker after truth." I doubt that physicists and cosmologists would be embarrassed. In their own way they are carrying on the ancient tradition of poet-seers and religious mystics. In common is a yearning for some ultimate lucid vision. They also share a sort of moral ambivalence, for by the very nature of their undertaking they must be modest and yet tempted by immodesty. Like religious aspirants, scientists easily slip into arrogance. Yet to attain truth, they must be humble; indeed, a scientific genius is often one who to an unusual degree shows charity to ideas and facts that others disdain to consider. Again, like religious aspirants, scientists are austere. Their workplace is austere, their schedule is unrelenting, their language is the dry bone of symbols and numbers. Yet the reality they discover is as splendid and strange as that of religious visionaries. It is this combination of extremes—austerity and splendor, the very small and the very large—that makes the scientific picture plausibly one that God may draw, and can give rise to the feeling in scientists that as they find their way into this reality they approach something that may as well be called God.

Awareness: The New Reality

When animals first emerged, they were a new reality in the universe, as also were their sensation and feeling, their cognition and experience—their world. What that world is like is unclear to humans and probably must remain so. This is true even of the world of our oldest companion, the domesticated dog. Large differences in sensory equipment and body build, as well as differences in the scope and power of the mind, separate us from the dog and most other beasts. Of course, there are similarities; there must be, if only because all animals have to know how to find their way about, recognize places that provide food, shelter, and possibilities for mating, as well as places of danger. At the abstract level of "paths and points," a human being may well be able to apprehend, if not exactly experience, another animal's world. This common world, its centers of high emotion notwithstanding, is deficient in particularity and vividness. It is schematic. For a human to say that animals live in such a world may sound like speciesist slander, yet just some such world is the day-to-day human reality. All too often we sleepwalk through life, turning right at this store, left at that tree, nodding to someone here, chatting with another person there, our sensorial and imaginative powers barely tapped.

What if these powers were fully tapped? What would reality be like then? Full use takes time—a lifetime—and is the true aim of education. Strange to think that the familiar path of learning might just conceivably raise one to transcendent heights. The path bears different meanings, from literal to figurative, depending on whether one has in mind the individual, the group, or humankind. An individual is able to learn in a precise sociopsychological sense. As his faculties develop, new things and new ways of looking emerge—new, though, only for him and not necessarily, or even likely, for others. Yet "unlikely" is not "impossible." An individual is, after all, unique; his new experience and discovery may be his alone and not sharable, or are sharable only at a future date, when the others have had time to catch up. As for the group, it too may be said to have a history of learning; its world opens up through the exercise of its collective intelligence. At the scale of humankind, "learning" becomes more a figure of speech than a precise psychological process. Indeed, as one surveys the millennia, one has the irresistible impression of punctuated emergence—periodic eruptions of the new into history, rather than of anything steadily cumulative and progressive.

"New" is a problematic word. What is discovered in science already exists or has existed. Discovery, then, is of the preexistent—the old; new is a quality of awareness in the discoverer. A homely analogy is that of waking up from sleep to see a brightly lit world as though for the first time. Another analogy is geographical exploration: An old continent is discovered, yet it is new to the explorer who understandably calls it such. There is, however, a difference between these two examples. Waking up doesn't need technical-material support; exploration does. Religious enlightenment is like waking up in that it doesn't need technical-material support; scientific enlightenment, by contrast, does. Nevertheless, religious enlightenment, or, more modestly, the sort of awakening that a good liberal education provides, though it may not require technology, does require technique and discipline (a cleansing of the mind's lenses), and cultural support—for example, inspiring models of what can be done. With scientific vision, the new depends not only on the above, but also, increasingly, on the availability of instruments, laboratories, and buildings designed to accommodate them. This vast material pile is itself a new reality in the world; in addition, it contributes to the discovery of new facts and the coming into existence of new qualities of awareness. How wonderful it is, Francis Crick reflects, that the fundamental building blocks of life, RNA and DNA, which existed for billions of years, became a fact of life's awareness of itself only forty-five years ago.[10]

Scientific awareness is wonderful; it is also highly specialized. As wonderful in a different way is the vast expansion of human appreciation of reality, not through the instruments of science, but through the education of the natural senses. Such education is best done by urging students to study the developmental paths and works of exceptionally talented individuals and groups.

Growing Up

Humans, however, become "students" the day they are born. What is the character of their learning curve? Is this curve much the same whatever a person's familial background and culture? The answer is yes. It is much the same in the sense that biological nature provides all humans with a common apprehension of reality. Whatever the differences between one individual or group and another, they are insignificant compared with those that exist between ourselves and other species, including our cousins the primates. That commonality among humans is especially evident in early childhood, for not only is the behavioral repertory of all infants much the same, so is their "world-view." Living in a rain forest or in a desert makes little difference, for infant views seldom extend beyond three feet; they can all be best described as tiny, unstable, and fragmented.[11] Divergence becomes more and more pronounced as, with the passing of the years, culture takes the lead to direct and refine nature's gifts. Even so, the upward curve of emotional-intellectual maturation has much in common cross-culturally. Everywhere, children's physical and worldly competence increases as they grow older. This gain is public; everyone can see it. Less observable, and also less certain as a necessary effect of maturation, is the gain in understanding of the real and the good.

Although maturation is desired and desirable, not all of it is gain. There are also losses—losses of innocence and of wonder. Loss of innocence is acceptable if knowledge and wisdom take its place. But loss of wonder and creativity? Surprisingly, only Westerners from the seventeenth century onward have learned to esteem both. Other people tend to see wonder and creativity, or (more prosaically) curiosity and probing, as an immature phase that is best left behind. On reflection this is not so surprising. Historically, children's constant need to know and find out was suspect, for if it developed at anything like the amazing pace of early childhood, it would be a constant threat to the stability of the social order. Westerners started to appreciate childhood at about the time when they were acquiring greater overall confidence in themselves, an effect of which was greater permissiveness toward the mind's playful tendencies at all stages of life.[12] Another reason for the difference in attitude toward childhood is this: In premodern societies, childhood may be filled with strange happenings, but so is adulthood; after the initiation rite, one simply moves from one kind of enchanted world to another. In modern societies, by contrast, the child's magic kingdom is left behind when one reaches a certain age. Adults, understandably, look back to it with a sense of loss.

But is this shift from enchantment to disenchantment inevitable? Is maturation necessarily a move from fantasy to the real? And is fantasy necessarily opposed to the real? One way to arrive at an answer is to look more closely at what can and often does happen in maturation. This means starting with children, and the first thing we notice about them is that they like to play. They like to pretend and live in a world of their own making; they pretend to be animals, they transform reality imaginatively so that a bookcase becomes a cliff, a broomstick a horse, an upturned chair a fortress. Educators now realize that children must be allowed to live at least some of the time in fantasy. Fantasy in children is not escape from reality but the natural means by which they explore and come to terms with it. Pretending to be animals is the quickest way to understand animal behavior, which has practical value. It is the only way to be emotionally engaged with them, see them as fellow creatures that share human needs, that are dependent, like humans, on the resources of the earth. Moreover, by pretending to be animals, children learn who they themselves are. Who am I? is one of the most difficult questions for humans to answer. Children make a good start when they see that they have the characteristics of an eagle, bold and fierce, or of a chicken, vulnerable and scared, that they are both and neither, that they already are somebody but can be somebody else.[13] Children, of course, also pretend to be the grownups they admire, thereby preparing them to be such grownups. Pretense is a technique for wakening skills and talents that would otherwise remain dormant. The psychologist Robert Hartley gives a suggestive example from his own life. As a youngster he showed a distinct lack of talent for writing. One day, when faced with yet another writing assignment, he suddenly thought of a person he had seen on television who had a way with words:

> It suddenly occurred to Hartley to write as though he were that accomplished writer, translating his own ideas into what he imagined was the expert's style. He might begin with "The man went into the room" but would realize his model would be more likely to write "The tall man went into the room." He continued translating his own words into those of the person he pretended to be, until he ended with something such as "The tall, thin man limped painfully into a [dimly lit] room." The experiment was a great success. Hartley's teacher was delighted at the remarkable progress in his skill, and from then on Hartley let his alter ego do his writing for him.[14]

At about the age of seven or eight, children begin to abandon pretense and fantasy in favor of realism. They have not lost the knack; they just prefer not to exercise it. They are more eager to be a part of the adult's world and do practical things, including those for economic gain, such as delivering the newspaper. They are also eager for social life. To participate in it, they must know how to communicate and so, in the interest of social talk and informational exchange, they forgo inventions that can have immediate meaning only to themselves in favor of the commonplace and the widely sharable—an objective world in which an armchair is an armchair and not some other thing.

Imaginative children thus turn into dull youngsters and duller adults. But is this inevitable? What does happen when one becomes an adult? In premodern times and places, the youngster, after an initiation rite, enters the adult's own world of magic and religion. For all the importance of this step in the eyes of its practitioners, I wonder whether it constitutes real growth, for real growth in a human person, as distinct from a plant or an animal, ought to contain within it the potential for further growth. Whereas children's fantasy is a genuine exploration into the real, an opening out to the world, adults' magico-religious customs, constructed against a background of anxiety and fear, may well be a clamping down, a terminus, a formalization of beliefs even more rigid than the houses they build.

Demonic spirits that induce feelings of dread and demand submissive, irrational behavior supposedly no longer afflict citizens of modern society. Is this true? The answer is a qualified yes. Much has been discarded as unworthy of an enlightened modern individual, yet primordial habits obstinately linger. Dr. Samuel Johnson and General Charles Gordon, noted for their common sense and practicality, avoided stepping on pavement joints for fear of incurring bad luck. I myself rise above that particular superstition; on the other hand, I still prefer to walk around rather than under the ladder. Charles Addams's spooky cartoons continue to make sense to me, as they do to his many admirers; driving past a cemetery on a dark night I resist glancing at the rearview mirror for fear of seeing someone who shouldn't be there. Modern high-rise hotels may not have a room number 13, or even a thirteenth floor. Ancient *feng-shui* techniques are still used to uncover sources of ill omen in brash, commerce-driven Asian cities such as Hong Kong and Singapore.[15]

These examples are easily multiplied. Many of the old ways of thinking remain, and they make for a certain charm if not taken too seriously. Not taking them too seriously is a sign of sophistication. As people mature, they ought to move from one kind of enchantment to another, from fear-driven fantasies to free yet disciplined exercises of the imagination. Imagination so exercised is actually quite commonplace, a fact we tend to overlook because our attention is too narrowly focused on exceptionally talented individuals and their works. We all know that Michelangelo was a genius. We may have heard it said that when he looked at a block of marble, he saw a figure crying out to be liberated. He chiseled away the marble prison, and the result was a magnificent work of art. That is magical! We are not Michelangelo; our imagination is not of that order. Still, even the humblest artisan, in a creative burst, accomplishes something similar: He or she looks at a stone and sees what is not yet there. That degree of imagination is necessary to making anything at all.

An Enchanted World for Rational Grownups

Premodern agricultural societies were more haunted by uncertainty than enchanted by the promises of nature, more weighed down by foreboding than buoyed up by delight. Is it possible to have the one and not the other, or more of the one and less of the other? I think so. Modern science has succeeded in liberating people from shadowy powers. At the same time, it has opened up a universe of mind-boggling majesty and strangeness. Isn't this improvement—progress?[16] Why, then, is there the widespread feeling that a pall of grayness has settled over the earth and that science is to blame? A number of reasons account for the heavy mood. One is that a long history of subservience to nature has made the freedom granted by science disorienting—an ambivalent good; we miss our former master, who, though often harsh, at least brought drama and the illusion of dialogue into life. The second reason points to a key trait of modern science, one that accounts for its success. This is reductionism—the exclusion of all information not strictly relevant to the problem at hand. It has encouraged a dismissive habit of mind captured by the expression "is nothing but." A table is nothing but a mass of whirling atoms, the sunset is nothing but a perceptual illusion. Most of us are not scientists, but we use technology, and we are well aware that technology is reductionist in regard to the human habitat; that is, it tends to thin and spread out its inchoate richness, flushing out nooks and corners in which mystery can flourish.

An analytic and reductionist approach to the study of matter has given people enormous power, which proves that it firmly engages the real and is not just an arbitrary procedure; one facet of the real, I should add, and not the real as such, which, if we follow Kant's position, is beyond human reach. Though we cannot know the real in itself, we can know more of its facets than the one that science importantly provides.[17] These other facets, both individually and in combination, are revealed as our mind and sensibility grow in compass and depth. In contrast to science, such growth comes about by a habit of mind that is more synthetic than analytic; it leads a person toward a richer, more evocative reality. I am saying, then, that if the past had poetry, so does the modern age, and if children inhabit enchanted worlds, so do adults, who get there not by reverting to magical thought but by developing the powers—I almost said the miraculous powers—of movement, language, sight, and hearing.

Movement: The Dance of Life

"There is a story told in *Treblinka* by Jean-Louis Steiner about a dancer standing naked in line waiting for her execution in the concentration camp. A guard tells her to step out of line and dance. She does, and carried away with her own authoritative action and with her individuality she dances up to the guard, takes his gun and shoots him."[18]

Talk about empowerment! Here are prisoners reduced to naked emaciated bipeds, utterly without human dignity and will, waiting in line to be disposed of like garbage. Guards stand by and watch, chatting to pass the time and making salacious comments on the naked bodies. As a last humiliation, a guard picks out the dancer and orders her to dance. One can imagine the awkwardness of the initial movements, and one can imagine the soldiers snickering. Then something remarkable happens. The dancer picks up speed and confidence, and as she does so, the dead air around her begins to stir; the place itself, and not just the dancer, comes to life. When she takes the gun from the guard and shoots, that blast is the blast of life.

Sound is life. Movement is life. Children after school rush into the field, shouting. A few in sheer exuberance turn somersaults. Younger children tumble over the monkey bars, their beaming faces sometimes up, more often down. One day through a restaurant window I see a father with his son on the opposite side of the street, stopping by a pet shop. The boy wants to be lifted up so that he can see better. What *I* see from my window seat is a remarkable pas de deux, performed without the least self-consciousness. The boy seems not so much to be lifted up as to rise up, with just the slightest assist, to the height his father's shoulder. As one arm goes around his father's neck, the other arm rises to point at something. A moment later, the boy doesn't so much clamber down as slide down in one sweeping movement.

If one cares to look, there is a great deal of physical, gestural elegance and daring that charges the atmosphere, "enchanting" it, in the schoolyard, in the dusty lots where children bounce balls and—yes, annoying as it can be—on curbs, over which boys with turned-around caps leap as though their skateboards were magnetized to their feet. Young people move with ease, which is a kind of grace, and even the gawkiness of adolescents projects a certain vulnerable charm. What happens when they grow up?

And what happens when a society moves from traditional to modern? Perhaps these two questions have a common answer.

Ritual and dance were inseparable in folk and traditional societies. Very likely they had a fervency and seriousness that the choreographers and dancers of modern times cannot quite replicate. They cannot be replicated because the extreme situations that prompted the old dances and made them necessary have largely disappeared. What extreme situations? Often a crisis of nature, such as a shortage of game, wilting crops, lack of rain, too much rain. The Pueblo Indian corn dance and rain dance both promote natural abundance. They are wonderful dances to watch today, hypnotic and inspiring. The whole plaza takes on rhythmic life, its dusty surface becoming the taut skin of a vast drum, played on by the rhythmic songs and pounding feet of the dancers. And yet something is missing in the modern versions. The movements may be as practiced as ever, the decorations correct to the last detail, and improvements may even have been added. Missing are the anxiety and the hauntedness. No matter how bad the corn crop, starvation in the late twentieth century is no longer a serious threat. Death, a presence casting its pall of unease over every ritual and drama in earlier times, has withdrawn. What remains in the corn dance—in the Yaqui deer dance, the Australian emu dance, the English Maypole dance—is the enchantment of art.[19]

In the modern world, as children grow up they lose certain natural graces of movement peculiar to their age and acquire the behaviors and gestures appropriate to adulthood, which, while serviceable, are lacking in the sort of elegance assumed matter-of-factly by adults of "good society" not so long ago. The slight bow, the hint of a curtsy, the tipping of the hat, the flourish of the cigarette holder, the dance at the door ("After you!" "No, after you!") have all become museum pieces. We are more than ever dependent on professionals to remind us of the beauty of movement, its power to transform—enchant—space. Many years ago I watched a film of Margot Fonteyn performing in *Romeo and Ju-*

liet. The scene that dwells in my mind is a sequence in which Juliet teases her old nurse, who chases her charge around the room. Juliet runs, dodges, floats on and over the bed. Fonteyn must have been in her forties then, playing the lithe and pliant teenager. But no real teenager could have been more lithe, more fluid in her movements than Fonteyn. She seemed to float. I was so transfixed by the movements of her arms and legs, torso, neck, and head—creating an arabesque that magically stays in the air—that I don't remember that her feet ever touched ground.

Language: The Creation of an Enchanted World

Movement is the animal universal; all animals move. Language is the human universal; all humans speak. Children move with a poignant grace that they may never regain in adulthood, unless they become athletes or dancers. With the use of language too children show a remarkable early flair that they appear to lose later. Parents and educators have often observed that young children like to play with language and often come up with striking metaphors. The psychologist Howard Gardner provides the following examples: To one four-year old child, a streak of skywriting is "a scar in the sky"; to another of about the same age, her naked body is "barefoot all over"; to a third, a flashlight battery is "a sleeping bag all rolled up and ready to go to a friend's house"; and so on. In play, a pillow is a baby, a banana leaf a houseboat. Much of the magic of childhood is expressed by, and owed to, these metaphors and metaphorical—that is, transformational—ways of seeing. Gardner notes that they are either perceptual or based on a similarity in action: A flashlight battery looks like a rolled-up sleeping bag; a banana leaf does not look like a houseboat, but in play it works like one. What is beyond most young children is the psychological figure of speech—for example, a heart of stone, or Shakespeare's famous comparison of love to a summer's day. Confronted by "heart of stone," a child suddenly turns realist and may see a guy whose heart is actually made of stone. As for Shakespeare's comparison, it would seem to the child far-fetched or just plain incomprehensible.[30] Yet an adult understands these figures intuitively; indeed, there is no other way, for logic will never persuade anyone to see the aptness of bringing these words and concepts together.

So who now is steeped in fantasy? Adults. But fantasy—or imagination, to use a more approbative term—in these instances does not end in a private, dreamlike state, unsharable with others; rather, it ends in a new, more expressive reality that can be shared, one in which love's ineffable character is made public by being coupled with a summer's day. The coupling is a novelty—an invention. Yet it seems unstrained, natural. Someone who comes to it for the first time may even be struck by its inevitability, as though it existed in the nature of things—inside the brain and out there—and had simply been *discovered* by the poet in a flash of inspiration.

Let me take up another difference between children and adults. Young children often see significance in the particular—a daisy, for example—that adults fail to see. This loss of wonder can, however, be regained through the devices of art. A daisy that is just one among thousands is invisible. It becomes visible, even worthy of admiration, when someone makes the effort to isolate it by removing it from a crowded field, putting it in a little vase, and setting that vase on a pedestal. Children do not have to go through these steps, for they have a natural tendency to see objects as isolates. Their world is vivid and aesthetic, made up of attention-grabbing particulars, rather than gray and practical, constituted by means and ends such that the ends themselves quickly become means and so lose their claim to sustained regard. Adults are not altogether bereft, for, as we have just noted, they can recreate vivid particulars by artistic means. They can also do so in a way not normally available to the child, the way of context, for though context can diminish meaning, it can also enhance it. Consider a poem that the sixty-four-year-old William Wordsworth scribbled in a child's album. The poem praises service:

> Small service is true service while it lasts:
> Of humblest Friends, bright Creature! scorn not one:
> The Daisy, by the shadow that it casts,
> Protects the lingering dew-drop from the sun.[21]

Placed between sun and dewdrop, the daisy is transformed into a luminous being that can cast a spell on child and adult alike. Not, however, necessarily the same spell; the adult rather than the child will be affected by the spell that depends on the richness of experience and context. A question now nags to be heard: Does power emanate from the daisy, or are we witnessing here a purely subjective emotion in the sophisticated adult observer? This way of posing the question excludes so much that it sounds more like an academic parlor game than a serious search for understanding. At issue is not a simple phenomenon that is encompassable by a question in the form of "either/or"; rather, it is something of great complexity and subtlety, involving different kinds of objects and events, both physical and psychological, that would take pages even to describe, much less explain. Perhaps only a poem—Wordsworth's poem, for instance—can hint at its scope. Still, if I had to state in one sentence just what is at issue—what is it that we are being asked to attend to and understand—I would say this: It is the conjoint presence of sun, daisy, dewdrop, and human person, a body overwhelmingly massive and powerful, bodies touchingly delicate and transient, their physical as well as aesthetic-moral relation to one another, at a particular moment in the evolutionary history of the earth.[22]

Sight: Visual Sophistication and Landscape

Children like to draw. When very young, they like to cover a blank page—or their bedroom wall—with bold strokes. Are they already young artists? Well, not quite, for what they really enjoy is not so much aesthetic creation as their own effectiveness in making some kind of impact. By about the age of five or six, children abandon these abstractions, which adults so admire, in favor of realistic depictions of tree, house, man, woman—"things out there." They move out of their egocentric shells in order to participate in a common reality. They seem to know that to communicate effectively with others, they must learn how to draw objects that others can easily recognize. This realism, as we have noted earlier, also applies to language. Language, of course, is something children must continue to use, whereas drawing pictures can be abandoned without social cost. In continuing to use language, older children, then adults, enter smoothly into the enchanted world of psychological metaphors; they appreciate innovative speech in others and strive to avoid clichés themselves. But few children, as they move into adulthood, continue to draw and paint. This does not mean that they never acquire aesthetic discernment—never learn to see beauty "out there." Quite the contrary. Children may give up drawing and painting precisely because by a certain age they learn to see for the first time truth, beauty, and goodness in nature and in human artifacts; and they realize, also for the first time, forcefully, that they are far from possessing the technique to do them justice. The abandoned paintbrush is an act of homage.[23]

Suggested here is a progression from the egocentric and often fantastic world of the very young to the realism of the young and finally to the enchanted world of teenagers and adults. I am reversing conventional wisdom. I am saying that even in enchantment, maturation is more gain than loss. Let me continue this line of thought with additional evidence. Consider the psychologist S. Honkaavara's findings in regard to how five- to six-year-olds understand landscape painting. Now, when adults are shown landscapes, they readily say of them that this one is "happy," that one is "gloomy" or "sad." Young children, by contrast, find such attributions baffling. It is a different matter when they are emotionally stirred. Happily playing, they have no trouble acknowledging that their playground is happy; if they are insecure and frightened, the landscape turns into an ogre.[24] Projecting a mood onto reality comes naturally to human beings, adult and young. The older human being's distinctive achievement is to see mood in the world that is independent of his own mood. Feeling sad, he goes outdoors and sees a woodland dappled in sunshine—a happy scene that has the power to change him and make him happy. Or imagine a woman in a serene mood strolling into a cove. Rather than having her feeling projected onto it, it, perhaps because a passing cloud has cast it into shade, acquires a menacing mien that overpowers her. One more example. A little boy, tired of his tricycle, abandons it on the sidewalk. The father retrieves it. He sees the tricycle all by itself, turned on its side, a look of forlornness emanating from its (momentarily) useless big wheels. The realist child has gone to bed, but the father stays awake in his—the adult's—enchanted world.

"Landscape" has a curious significance for human beings. The word itself is heartwarming, like "home," but with a cooler tone. One may think landscape a common and even universal way of perceiving and experiencing, but this is not the case. Young children have no patience with it, as parents know when they stop the car to admire a view. Among the world's many cultures, only two—European and Chinese—have independently made landscape a major genre of art.[25] So we have to conclude that, far from being universal, it is a specialized way of seeing. Yet it is a way of seeing that, once discovered and made known, quickly finds admirers and adopters. America, daughter of Europe, developed its own superb landscape art. Japan, pupil of China, in time created landscapes—for example, Hokusai's scenes of Mount Fuji—of outstanding originality. Still, Aboriginal Australia offers the most convincing example of landscape's natural appeal, for its culture is radically different from that of both Europe and China. Aboriginal Australia has produced very talented landscape artists. Their works are much sought after by tourists and connoisseurs alike. Looking at their scenes of sun-soaked boulders and wraithlike eucalyptuses, it is hard to remember that landscape painting is alien to their tradition, that it is an import from Europe which they have perfected and made their own. Without doubt, landscape painting is immediately comprehensible and admired in almost every part of the world today. People who speak different tongues and cannot communicate can nevertheless understand one another as they nod appreciatively over pictures of a landscape. This widespread appreciation, well established by the middle of the twentieth century, is owed to the influence of paintings as they are displayed in galleries and as postcards, but far more to the influence of the camera, which, once it became cheap enough to buy at the local store and easy enough to operate, became the inseparable companion of every tourist. The millions of cameras clicking away on scenic routes, the seaside, all sorts of parks and wilderness areas, are the most prolix generator of a picturesque conception of reality.

What gives landscape this extra measure of importance, making it more than just a local predilection like that of Wisconsinites for cheese? Jay Appleton has consistently argued that when landscape is considered in its two principal components of "refuge" and "prospect," its role in survival is immediately evident. So perhaps this is the clue: survival. From a safe corner one surveys the terrain ahead, taking stock of its dangers and opportunities.[26]

Using a slightly different language, I say that landscape is made up of "place" and "space," the place of stability and confinement and the space of vulnerability and freedom. Some of life's fundamental polarities are thus presented.[27] But survival does not by any means exhaust landscape's appeal. Aesthetics is a factor too. Aesthetically, landscape satisfies a human need for harmonious resolution between such basic binaries of human experience as vertical and horizontal, foreground and background, illumination and darkness. Furthermore, in landscape, people find deep satisfaction in being both attached and detached, for landscape is neither embeddedness in locality nor a God's-eye view of the world but a position somewhere in between. From that position one can see and be sympathetic to human undertakings and human fate, yet not be totally involved. Total involvement may sometimes be necessary, but it is not always desirable, for it usually means the loss of the ability to contemplate and reflect, to disengage oneself—to escape.[28]

Hearing: The Mystery of Music

Heavenly music—the two words have a natural affinity. Music that is not heavenly is not true to its calling, and heaven without music would be, one feels, static—dead.[29] Whenever one tries to *visualize* heaven, one comes up with rather pedestrian pictures—bejeweled city, garden, rolling pasture—that are, if one pauses to reflect, quite implausible. Music in heaven is different. True, the harp has become a cliché, but to a lover of music, the thought of a Bach fugue or a Mozart sonata in heaven is not out of the question. Because sound is not something one can reach out and touch, and invokes neither a clear picture nor a clear story line, it seems to belong to the Other World. However, for those same reasons, many would consider a heaven of pure music quite unreal—the real, for them, being thoroughly bound up with the senses of sight and touch. Touch is the ultimate test of the real, followed by seeing, followed—at a considerable distance—by hearing. Eyewitness can be trusted, hearsay not. Eyes can deceive—what one sees is a mirage—but "hearing voices" is classic hallucination. Sadly, that great conception of the cosmos as a vast musical instrument, filled with rapturous sound, happens not to be true. Originating with the Greeks and sustained by literary works (think of Shakespeare) through the subsequent periods of European history up to and including the Renaissance, the music of the spheres finally stopped playing in the seventeenth century, prompting Pascal to bemoan the eternal silence of outer space.[30]

Compared with pictorial art, music is mysterious. Much of pictorial art is, after all, representation. The artist tries to recapture the beauties of nature, often with a sense of not quite succeeding; the original at its best is elusively superior, whether this be a single apple, a human face, a landscape, or the abstract splendor of the aurora borealis.[31] The musician also tries to represent the sounds of nature—a bird's mating call, the babble of a stream—but these are mere incidents in a large composition for which there is no parallel in the external world. Mont Sainte-Victoire is a match for Cézanne's painting, but nothing in nature can remotely match Beethoven's Pastoral Symphony. The Pastoral Symphony, in other words, is not discovery, not imitation, but almost pure invention, yet an invention that does not sound (to the music lover) in the least arbitrary or conventional. On the contrary, it sounds as though it has always existed—as though it is or ought to be one of the great Platonic forms. Actually, bringing in the heavy gun of the Pastoral Symphony is quite unnecessary, for just about any musician's melodious tune is a decisive improvement over nature's normally discordant notes. Perhaps for this reason music is widely regarded as the exemplary paradigm of harmony—especially social harmony, which, however, reaches perfection only in heaven, where music reigns supreme.

Words can be spoken monotone—from the throat up, as it

were—accompanied by hardly any facial or bodily gesture. Singing, however, draws on deeper resources of the body and is a greater commitment of the whole person; the face is highly expressive, and if singers do not gyrate conspicuously it is because, as Jacques Barzun puts it, their visceral impulses—their need to move—are already strenuously exercised "through their lungs, throat, and diaphragm."[32] Group singing powerfully consolidates community—the human community first of all, but reaching far beyond it to nature and to the material objects important to human well-being. That is a bare-bones statement. Precisely what the satisfactions are is harder to say. A singer is keenly aware of her own body as she sings yet may feel as though she were merging with other bodies, other singers, dissolving into a globe of vibrant sound. Reflecting on the experience, she will have to admit that it is "bracketed" from workaday life. But it may feel more real than workaday life and is certainly more memorable. Interestingly, as community singing decays in modern times, the image of heaven as a place of choirs where people do little else but sing—witness the cartoons—comes to the fore.

All humans sing, just as all humans speak. Instrumental music, though ancient and widely practiced, is not quite so universal. Yet one may well think that it ought to be—that people without some kind of musical instrument must have lost an art—so deep by now is the link between instrumental music, if only of the simplest kind, and human culture. Instrumental music extends and complements voice. It is felt to be animate; it speaks. But what does it say? A song, because of its words, delivers an understandable and fairly specific message, but what are we to make of the sound that emerges from an instrument? The temptation to come up with something like "water flowing over a weir" or "a woman sorrowing for her departed lover," based on the most far-fetched interpretations, is nearly irresistible. The musician's facial expression and body movement provide extra cues; an arched eyebrow conveys questioning defiance, a drooping shoulder resignation to fate. It goes without saying that these interpretations are utterly inadequate to what receptive listeners register under music's spell—moods, the range, precision, and delicacy of which take them by surprise into worlds that are sadder, more elusive, or grander and nobler than any they have known or can otherwise know.[33]

In earlier times music was just one component in a festive occasion, the others being words, which might soar to oratory, and bodily gestures, which might become dance. The meaning of the music—for that matter, the meaning of the entire ceremony—was taken for granted. It was not an event to be explicated or appraised. Multimodal activity, with emotion at full throttle, prevented in any case the necessary distancing and reflection. Even when the atmosphere was not all-involving and people might come in and out of a ceremony as they wished, there was no call for critical evaluation of a happening that so effectively distracted and absorbed. In our own time, popular music tends to combine instrumental music with song and dance, and possibly bits of narration. Hard rock not only has this multimodal character, it is also emotion-drenched; and even if, unlike in earlier times, there is a separate audience, that audience doesn't stand aside but is drawn into the performance by the booming sound, the flashing strobe lights, and the inviting gyrations (often sexual) of the singer-actors. The one immense satisfaction, common to these differing kinds of experience, is the loss of self in a large, potent whole, the small vulnerable body with its ineffectual voice absorbed into a great Body that hums and surges with power.

A curious phenomenon, unique to the West, emerged sometime near the end of the sixteenth century. This was pure music or "music alone," as Peter Kivy puts it.[34] Music is "pure" when it is not mixed with gesture, dance, or song—especially not with song, for it uses words that have their own denotative and connotative meanings. Pure music is thus unaccompanied instrumental music. Its performance calls for a clear separation between musician(s) and audience. The idea that an audience should sit apart and do nothing but listen, that no member should feel free to get up and move around during the performance, is alien to premodern practice.

By the eighteenth century, listening to instrumental music in an especially designated or designed area had become the thing to do, as of course it still is. Such a shift in attitude could only happen if certain changes in society's nonmusical institutions had also taken place. The one word to characterize all the changes, interlocking and overlapping with one another, is "modernity." In music, around 1700, a new way of responding, a new understanding of what it means to communicate, a new sense of oneness, came into existence that resembled ideas that had emerged earlier with the rise of landscape painting in visual art. Landscape painting is like "music alone" in that it too requires a degree of distancing that is a key aesthetic-moral element of modernity. Instead of being immersed in an environment, people stand aside to become observers. Landscape art encourages pure vision as distinct from multimodal perception. In music, likewise, an audience forgoes the sound womb to face a spacious soundscape "out there." A dynamic range that has increased sharply since 1600 creates an illusion of foreground and background—deep, booming notes providing the one, rising high notes the other.[35]

Separation is prelude to union; if no separation—that is, no real individuals—then no real union. The pianist is on an illuminated stage, the audience is in a darkened hall. Both are alone. The pianist is obviously alone, but so, if less obviously, is each member of the audience alone in the darkened hall. The one plays, the other listens. Proper listening is not passive. Rather, it calls for the suppression of self and active attentiveness. No wonder few people do it well. Communal chitchat, I noted earlier, requires little listening. Likewise a traditional festival that features song and dance, for such an event has no separate audience. Almost every adult has a role, and even those who do not may have some other thing to do—minding a small child, shooing away a dog, chatting with a neighbor—which means that no one is engaged with the overall musical effect, with a created entity "out there." If there is something admirable in attending closely to that which is "out there," beyond the pressing needs and concerns of the moment, then the modern age contributes to that admirable trait by providing occasions when people strive to listen, and to do so not just for a brief minute or two but for long stretches of time.

What happens to the audience when a Beethoven sonata is in progress? No doubt individuals show striking differences in behavior and degree of attentiveness. Some snore, a few are absorbed in their neighbors or the chandelier, one or two may wonder whether the stock market will crash next week, and so on. That is only to be expected. We are like that. On the other hand, in redemption of such human foibles are the virtue of self-forgetfulness and the reward of total engagement with a type of music that may be literally meaningless but that to devoted listeners is meaning incarnate—sweatily passionate, coolly intellectual, cozily familiar, galactically remote. It is the epitome of order, yet studded with surprise, bound to place and time, yet somehow timeless and universal; perhaps most wondrously of all, it is sound that gives point to silence.

Invention and Discovery: Can't It Be Both?

After existing for billions of years, life suddenly discovers that it is made up of extremely long macromolecules. New awareness of this kind is fairly common in science. There must be something about the human brain—this bit of organic nature welling up at the tip of an otherwise unremarkable stem—that permits such self-knowledge. It doesn't have to happen; culture doesn't have to take that turn and indeed hasn't except in a small corner of the earth at a particular moment in history. Likewise appreciation of pure music. It is a new way of relating and being, a novelty in the universe. It doesn't have to happen and indeed hasn't except in a small corner of the world in the seventeenth century. By now, however, pure music, like a knowledge of DNA, is human heritage, accessible to all who wish to make it their own.

Pure music is not only an invention, it is also a sort of discovery; one part of nature has "discovered" a nonlexical, musical way to connect feelingly with other parts of nature. Discovery is of something that already exists. Pure music doesn't already exist in the same sense that molecular structure already exists. It is possibility become actuality, a new state of awareness, to be compared with sophisticated biological science rather than with DNA. A closer analogy, however, is with architecture, which, like music but unlike organic chemistry, is simultaneously a construction and a revelation of our own deepest emotional nature. Architecture has been called "frozen music." The possibility for it has always been there, but there has not always been architecture. Eventually, beautiful edifices are built. In such an edifice, a human being may feel that a profound yearning of her nature has been met, that by entering it she has come into her own. Great music has a similar effect. Even as one agrees with the idea that listening to Bach is escape from the messy realities of life, one may still feel deep down that this is not so, that, on the contrary, listening to his great works is to feel that one is entering a magnificent building that is also, strangely, one's intimate and rightful home.

Human Relations

"We are nothing but meat!" the British artist Francis Bacon says angrily. The anger comes out of frustration with his fellow humans who, with incurable childishness, are forever decking themselves in qualities that they believe entitle them to esteem.[36] Boastful claims to specialness notwithstanding, in actual behavior people have all too often treated one another as of little or no consequence. History is a vomitus of atrocities; our own time has produced the massive genocides of Germany, Cambodia, and Rwanda. It is black with exploitation; people are there to be used for someone else's pleasure and profit, and this was so not only in despotic empires but also in democracies—in ancient Athens, which depended on slave labor, and in modern democracies that depend on the labor of migrants, aliens, the disadvantaged.

Of course respect is given, but tautologically to persons of power and prestige; affection flows, but only among one's own family, kinfolk, and neighbors; human dignity is recognized, but foremost—or even exclusively—in one's own group. It is common for people to have a special word, an honorific term, for themselves, implying their full human stature, and some lesser word for others.[37] America's current distrust of centralized government is itself an illustration of this deeply ingrained way of thinking. Will Washington bureaucrats put themselves out for stranger-citizens thousands of miles away? The question assumes that they won't, but that, on the other hand, local officials *will* put themselves out for their own community.[38] Taking care of number one, whether the "one" be an individual or a collective self, is a fact; thinking well of self and indifferently or ill of others is another fact; accepting facts and being guided by them is realism. Anyone who says otherwise can be dismissed as a fool peddling escapist dreams. Yet curiously, some of the world's greatest thinkers have "peddled such dreams." High claims have been made for humankind in Stoicism, Judaism, Buddhism, and Confucianism, all of which emerged on the world stage rather suddenly between 600 and 300 B.C. These universal religions and philosophies say that enlightenment and salvation are possible to all who seek it. If so, then all are put on an equal footing in the only things that matter.

Let us look at two such universal belief systems, Confucianism and Judeo-Christianity, which have roots in what Karl Jaspers calls the Axial Period.[39] They share certain values but also strongly differ. Does the one or the other, or both, have the ring of truth? Do they capture idealized pictures of human relations that, however contrary to local practice, stir immortal longings?

Confucianism

Confucianism is far less radical than Judeo-Christianity. For one, Confucianism accepts hierarchy in both nature and society; in its view the two are closely meshed, components in the one corresponding to and merging with those in the other. For another, there is in Confucianism a persistent deference to social status; respect, submission, and piety, though no doubt good etiquette for everybody, are directed primarily at the lower ranks. Third, family is the model for society all the way up to the empire and beyond to the cosmos itself, and family is hierarchical in its very nature. The family model encourages deference to superiors—to the local magistrate and, ultimately, the emperor as father figures. Enlightened paternalism is the result even when the model works to perfection. On the other hand, Confucianism has also sought, ambiguously and contradictorily, to depart from this traditional world-view. It has come up with ideas that appeal to the modern liberal democrat. For example, Confucius is reported to have said that "by nature, men are nearly alike and become different only as a result of practice."[40] They are alike in simple goodness, alike in possessing an ability to learn. Through self-discipline and education, ordinary men can become "superior men." The term *chün-tzu*, translated into English as "superior man" or "gentleman," originally stood for someone of high social position, but in Confucianist usage it increasingly takes on the meaning of moral rectitude, gentleness, and courtesy, qualities that anyone can acquire. Confucians choose to esteem achieved rather than inherited status. They seek to make *chün-tzu* mean, first and foremost, a moral being. But here is the rub: Removing inherited inequality can result in the creation of a meritocracy, another type of inequality, with not so much the diligently studious but the naturally talented on top. Some people seem to have an intuitive understanding of good conduct and moral behavior, others can only acquire it by application, and many just don't get it even when they try hard. Confucius's own words are harsh: "Those who are born knowing it are the highest. Those who know it through learning are next and those who toil painfully but cannot learn it make up [the bulk] of the people."[41]

A meritocracy would not, however, have disconcerted Confucius and Confucians. What's wrong with having the best people on top, provided, of course, that they *are* the best—best in administrative skills and, more important, best morally? Confucians know that the elite, even if they have risen to their responsible positions by legitimate means, can lose their moral anchor to start on paths of unbridled greed in the pursuit of sensual pleasure, wealth, and fame, and so turn into corrupt rulers and officials of insufferable arrogance. To Confucians, such excesses are the core of moral evil. They never tire of preaching against it, which in effect means denouncing those in power. So a sort of balance is restored. If Confucians urge respect and submission on common folk, they also urge restraint and obligation on officials and the upper class. Moral evil is a serious temptation only to the privileged. Farmers and artisans, for lack of opportunities to indulge, are better able to retain their simple goodness *(chih)*. Simple goodness is, however, a wobbly virtue that can turn into crudity and boorishness unless it is refined into cultivated behavior *(wen)*. A credit to the Chinese social system is that movement up from *chih* to *wen*, by means of education, is open to people of modest socioeconomic means. Cultivated behavior is highly desirable—

the Confucians have always favored civilization—but it too has its risks. It can become just slick smoothness.[42]

If I had to say in a sentence what is best in Confucianism, I would say that it presents human life and relations as a sort of holy dance, analogous to the cosmic dance that is discernible in heaven. Now, the word "dance" may be misleading, for some dances are full of frenzy, and that would not be to the Confucian taste. What Confucius and his followers have in mind is a stately affair, but a stateliness that is enlivened by genuine feeling. Two key words in Confucianism are *jen* and *li;* together they make the dance both possible and lively. *Jen* is the natural affection of people for one another. It is exhibited in all kinds of human relations, but outstandingly in those between parents and children, old and young. *Li* originally meant "sacrifice" or "sacrificial ritual," but it has come to mean appropriate human behavior as such, everything from court rites and diplomatic ceremonies to the most frequent and ordinary encounters in the home and on the street. Ritual and ceremony invite a certain rigidity if not pomposity—unfortunate tendencies that are to be shunned, Confucians repeatedly say. A rite is not only worthless if it is done without appropriate emotion, it also cannot be worth much if it can be performed only with difficulty, for a difficult course of action carries the sense of forcing, and what is forced is unnatural. A ceremony marks a high occasion. The rather elaborate gestures and steps that occur are *natural* to it; hence, provided that one is in a proper frame of mind, they should not be difficult to learn. In any case, anything less stately would go against the human grain. Confucians seem to say, paradoxically, that it is natural for people to acquire habits of consideration and courtesy; and that these are to be applied to all occasions, high and low. Even the casual greeting of a friend should convey, gracefully, a feeling of pleasure and regard. Habits may have to be learned step by step, but once in place they allow one to act without thinking, naturally.[43]

Two aspects of the holy dance are especially worthy of note. One is that respect is bestowed on another in the civil gesture. The issue is not whether the person *deserves* respect, or why anyone not socially important should be treated courteously—should be a part of the inclusive dance. The second point is that though Confucians are not against lightheartedness—Confucius himself shows a touch of frivolity when he sings in the rain—they stress that human beings and human relations have about them an aura of seriousness because they partake of the holy, a stillness that emanates power and virtue, supremely manifested in the polar star and the emperor on his throne, but also present or potentially present in every thoughtful individual.

Judeo-Christianity

Confucianism is a philosophy of virtue and human relations, rationally argued but backed ultimately by a sense of the sacred. Contemporary Greek thinkers (outstandingly, Socrates) might have found it sympathetic, for they too had a philosophy of virtue (the good) and human relations, rationally argued but backed ultimately by an inspiring conception of transcendence. Judeo-Christianity differs in that its God preempts the front stage; human relations and everything else are in the shadow of, or follow from, the fact of God's dominion. What fantasy is this? one may ask. Most people, past and present, have no use for it. Realistically, they embrace the tangible and necessary things of life and expand them to include deities and spirits only when forced to save appearance and to make life more operational—more subject to control in a mechanistic way. Why impose on the more or less manageable world a transcendent God who is too far beyond human powers to envisage and comprehend? To make divinity even more elusive, this high and abstract Holy (Wholly) Other is also a tangible Being who walked with our first parents in the cool of the evening, who in the form of an angel wrestled with Jacob, who in the Christian myth was even a babe in a manger.

A transcendent and omnipotent God, by shedding light and exercising authority everywhere, has disenchanted the world, freed it from haunted grottoes and dark woods, magic, sorcery, dread. An enormous weight was thus lifted off human shoulders. But in the process an even greater weight was placed on them by an insistent perfectionism that would have been annihilating but for the continuing existence of laws and regulations, customs and practices, dispensations and indulgences. These have always been a part of religion; they were and are a help to anxious men and women. In Judeo-Christianity, however, they do not guarantee salvation. Following them to the letter, as rabbis and priests insisted and still insist, is insufficient; it does not ensure righteousness in God's eyes. In His eyes even a prescribed sacrifice can be an abomination if not done in the right spirit (Isaiah 1.12–14); the holiness of the Sabbath itself, so comforting in its authorized abstentions, still does not excuse man from responding to those in need (Luke 6.6–11).

Judeo-Christianity's demand that the poor and weak receive justice, that they not be despised, has become by now so familiar that perhaps we forget that such moral precepts were not, and still are not, everywhere accepted even in principle. Pagan Romans did not accept them, modern fascists do not accept them, the Mbuti Pygmies, for all the general warmth of their communal life, do not recognize them. In Judeo-Christianity, even the outsider and stranger must receive his or her due. The Hebrews are instructed, "You shall not leave your vineyard bare, neither shall you gather the fallen grapes of your vineyard, you shall leave them for the poor and the stranger"; since you yourselves were once "strangers in the Land of Egypt," you should have learned not to vex or oppress the alien in your midst (Leviticus 19.10,

33–34). As the Yahweh of Israel turned into the God of all nations, and then the God of every individual in every nation, the idea of providential care for all moved to the forefront, culminating in Jesus' sublime call for perfection:

> You have heard it said, You will love your neighbor and hate your enemy. But I say, Love your enemies, bless them that curse you, do good to them that hate you, and pray for them that spitefully use you and persecute you, that you may yet be the children of your Father who is in heaven: for He makes His sun to rise on the evil and on the good, and sends rain on the just and the unjust. For if you love them who love you, what is so special about that? Don't even the publicans do the same? And if you greet your brethren only, what do you do more than others? Don't even the publicans do so? Be therefore perfect, even as your Father who is in heaven is perfect. (Matthew 5.43–48)

The ethical extremism of this passage is too easily passed over. We forget that to reasonable men of other times and places—to Confucius and Aristotle, for example—it could seem the ravings of a madman.[44] Even if these ancients could have caught a glimpse of the teaching's grandeur, they would have considered it unrealistic—an unconscionable burden on human nature and good will. Their modern avatars too will shake their heads. Yet since the appearance of Christianity, it is hard not to be somewhat moved. People proud of their rationality or just plain common sense are not entirely deaf to its appeal.

Also extreme, from the standpoint of most moral positions known to the world, is the reversal of privilege and status. It is one thing to say that the rich should give to the poor, that the strong should help the weak, that good citizens in high positions should not despise those less fortunate, but quite another to say that somehow the weak and the lowly (even the socially deviant—

a woman of doubtful reputation, a thief) are to be envied, for they may turn out to be the true children of God. How that undermines the delicious Roman virtue of magnanimity—a virtue that only magnates can practice! Judaism's outstanding revolutionary, Jesus of Nazareth, seemed at times bent on putting everything on its head: The first shall be last, the last first. The poor rather than the rich are blessed. Children are wise in a way that even the wisest adults are not. The prodigal son rather than the good son, the repentant tax-collector rather than the righteous Pharisee, elicit affectionate indulgence from God.[45]

Another affront to good sense is that numbers don't always count. The good shepherd goes to look for the one stray sheep, leaving at risk the ninety-nine that did not stray. The whole must be prepared to sacrifice for the individual, rather than, as has always been assumed, the individual must sacrifice itself for the good of the whole. Who is this individual that is made so much of? Someone of exalted status like the king, or just anyone? It turns out that they are the same person. The following passage has not dulled with repetition:

> Then shall the King say to them on his right hand, Come, you who are blessed of my Father, inherit the kingdom prepared for you from the foundations of the world: For I was hungry and you gave me food; I was thirsty and you gave me drink; I was a stranger and you took me in; naked and you clothed me; I was sick and you visited me; I was in prison and you came to me. Then shall the good people answer him, saying, Lord, when did we see you hungry and feed you, thirsty and give you drink? When did we see you a stranger and take you in, naked and clothe you? When did we see you sick, or in prison, and come to you? And the King shall answer and say to them, truly I say to you, inasmuch as you have done it to one of the least of these my brothers, you have done it to me. (Matthew 25.34–40)

Reciprocity is at the core of moral behavior in premodern times. It extends even beyond humans to gods, ancestors, and spirits of nature. Its ethos is essentially practical, as is indicated in the formulaic prayer of the ancient Romans, *do ut des* (I give so that you might give). Jesus would certainly have favored reciprocity, but exchange that is for mutual benefit, far from being at the heart of morality, is in his view only a point of departure: "Why should God reward you if you love only the people who love you? Even the tax-collectors do that!" He also offered the advice that "when you give a lunch or a dinner, do not invite your friends or brothers or your relatives or your rich neighbors, for they will just return the hospitality. Rather when you give a dinner, invite the poor, the lame, and the blind, and you will be blessed, for they cannot pay you back" (Luke 14.12–14). Again I wonder, Would a Confucian approve? Would any good and sensible person operating within the realm of the possible approve? Yet breaking the circle of reciprocity is common practice in any modern society. All individuals in our time are or should be made aware of their awesome indebtedness to their predecessors for every benefit they enjoy, from good health and long life to intellectual stimulation. Surely when sick and in need of surgery, they can feel gratitude for all the inventions of medical science (just think of one small item among them: chloroform). Not only are the vast majority of these cultural benefactors dead, but they are also dead strangers, and there is no way of reciprocating. One can only pass on what gifts one has to others—other strangers, including those yet unborn. As for the community of the living, firefighters risk their lives saving strangers, dedicated teachers and welfare workers exhaust themselves with no expectation of return from their charges, and the list of such salaried good Samaritans—their salaries incommensurate with what they do—goes on and on. Of course, Jesus did not have this sort of secular "linear giving" in mind. He was trying to break the egoistical calculations of narrow reciprocity by drawing attention to reward from God. Nevertheless, his perspective on the moral is far closer to and seems to anticipate the needs and practices of modern times.

Confucius preached *li* (rite), and *li* is concerned first of all with outward behavior. One should act in a seemly fashion, courteously. True, Confucius also said that the right feeling should go with the act, but conformity between inner state and outward expression was not the message he underlined. Generally speaking, the premodern person was not much concerned with the inner self; that came later as the sense of individuality developed, reaching a peak in the soul searchings of such early modernists as Montaigne and Loyola. Judaic tradition, like that of the ancient Greeks, stressed outward behavior over inner feeling. A rabbi would have said that dreams cannot have equal standing with reality, that dreaming about committing adultery is not committing adultery.[46] But Jesus famously said, "I tell you: anyone who looks at a woman and wants to possess her is guilty of committing adultery with her in his heart" (Matthew 5.27). Paul, the disciple, carried this emphasis on the inner state to the point of saying, "Even if I give away everything and offer my body to be burnt, but have no love, it doesn't do any good" (1 Corinthians 13.3). The driving force behind this insistence on conformity between "inner" and "outer" was an extreme revulsion against hypocrisy, which implies that omniscient God can somehow be deceived. Of course, that is not the case: "Whatever is covered up will be uncovered, and every secret will be made known. So then, whatever you have said in the dark will be heard in broad daylight, and whatever you have whispered in private in a closed door will be shouted from the rooftops" (Luke 12.2–3).

Jesus often spoke in hyperbole, a famous example being forgiving one's brother seventy times seven times. His moral principles often went far beyond, even reversed, good sense. He "invented" them with an artist's freedom. Yet in time they have come to be accepted and even—in modest degree—practiced. They are not wholly impossible after all; they may even indicate what human beings really are, or are capable of becoming. In the Western world, and more and more so elsewhere, it is assumed that strict reciprocity, whether in reward or in punishment, is not right, not even natural, for it puts a stuntingly low ceiling on the moral heights to which people can aspire. In the West, privileging the "underdog" is a commonplace of thought, if not quite of practice: The prostitute has a heart of gold. The downtrodden proletariat has not only justice but virtue (selflessness, courage, honesty, etc.) on its side. The legal system, when it works, adjusts the scales so that the convicted is treated fairly from the belief that he or she, in some larger sense, is also a victim.[47] As for Jesus' condemnation of secrecy, his prediction that everything whispered in "smoke-filled rooms" shall be shouted from the rooftops, that must have seemed a wild dream in his time, yet it too has become a principle of public morality. Just how fantastic this call for openness is shows up when we realize that ever since Neolithic times, when houses were first built and walls raised, conducting affairs in secret has been both a necessity and an inevitable cause of suspiciousness.[48] It is not privacy itself that is objectionable; serious discussion of any kind calls for quiet, secluded space. Rather, it is secrecy, which, whatever its original purpose, easily breeds a habit of deception.

Fantasies that in time become actualities were not fantasies even when they first appeared, for even then, in the midst of disbelief and head-shaking, they planted a seed that later bore fruit. Indeed, I would say that locking oneself into common sense and the currently practicable promotes fantasy by denying encoded potential, as though a seed could never become a tree, a caterpillar never a butterfly.

"If the Doors of Perception Were Cleansed . . . "

Often we walk through life with a figurative head cold such that the world around us seems a blur, and if we draw what we see, it comes out looking like a small child's picture, with trees that are all the same, their trunks straight, their foliage without individual leaves, and if there are flowers, they are all daisies. Other times—more rare—the world is sharply etched, brilliantly colored, dewily fresh as though a rain shower had just passed by, and solidly present, incorruptible. Thomas Traherne in the seventeenth century must have had some such experience, raised to mystical heights, for he wrote, "The corn was orient and immortal wheat which never should be reaped nor was ever sown. I thought it had stood from everlasting to everlasting. The dust and stones of the street were as precious as gold: the gates were at first the end of the world. The green trees when I saw them first through one of the gates transported and ravished me; their sweetness and unusual beauty made my heart leap, and almost mad with ecstasy, they were such strange and wonderful things."[49]

What must one do to see the world this way? Certain chemicals help. Some American Indians, for example, eat the root of peyote in their sacred rites. The rites themselves heighten awareness, and the active element in peyote, mescaline, heightens it further. In modern secular times, more than ever people depend on chemical stimulants to dispel the "head cold" of daily life, even if it is just the caffeine in the morning cup of coffee. Curiously, although we know what drugs can do, forgotten or just out of mind is the power of technology to effect this shift, dramatically as after the successful removal of cataract, and routinely each time one puts on a pair of spectacles. Here I am, groping my way through a forest of blurry objects; I put on my spectacles, and presto, reality! Whereas in an earlier time each person lived in a world individualized by the character and degree of his or her defective vision, we now, thanks to this ingenious tool, live in a world that is not only more vivid but also shared.

Art, historically, has had the role of lifting life from the doldrums. Under its guidance and stimulation, things and events that have gone flat take on new savor and sparkle. Realistic art, no less than fantasy, has this power, but realism is at constant risk of assuming the dung color of that which it depicts, becoming a dull pictorial replication or pedestrian journalism that cannot invigorate. Resorting to fantasy may not be the answer either, for the world it presents is too remote from common experience. But is this necessarily true? Consider two genres of literary art—fairy tale and science fiction—that found many admirers in the nineteenth and twentieth centuries. Both are full of marvels; both take the reader as far as possible from life's humdrum realities. Important differences, however, exist. One is that whereas the marvelous in science fiction is the unexpected and bizarre, the marvelous in a typical modern fairy tale is not only that but also such familiar objects as a loaf of bread or a tree stump and such "ordinary" people as milkmaid and woodcutter; and it achieves this effect by placing them in a world that hints at vast panoplies of space and time—exhilarating possibilities of existence. Another difference is that fairy tales, as distinct from works of science fiction, satisfy a profound human longing for communion with, and not just control over, living things—and beyond them, rock, air, water, and fire.[50] In a sense, language as such, since it is disposed to animate every kind of creature—the mountain has a "face," the table has "legs," the wind "howls"—already makes communication and communion with them and among them seem plausible. But in a fairy tale, such kinship is explicitly recognized, and it remains for the storyteller to reveal his or her talent by working out subtle differences among the richly varied categories of being. Fairy tales, again unlike science fiction, tend to be more strongly moral in character; conduct there is viewed through an ethical lens, heroism being a matter of helping others rather than daredeviltry.[51] Whereas science fiction may well feature good guys and bad guys, the genre is not designed to explore the nature of good and evil, as it is in fables as different as Tolstoy's "What Men Live By" (1881), Oscar Wilde's "The Happy Prince" (1888), and J. R. R. Tolkien's *The Lord of the Rings* (1954–55).[52]

Relatedness and Nothingness

The temptation of "cleansed vision" is to make idols of objects that have suddenly taken on import, or to value the ecstatic experience itself to the exclusion of everything else; if one doesn't worship, one wants to merge. Relatedness, other than this adulation and coupling, is conspicuous by its absence. The bright objects themselves stand alone, each demanding total attention. As Aldous Huxley put it, "I saw the books [their individual being and meaning], but was not at all concerned with their positions in space."[53] Drugs that produce sensations of orgasmic power and visions of mystical intensity do not turn their consumers into better, more enlightened people. One reason why they do not—apart from the chemical damage they inflict on the human system—is this fixation on unique particulars at the expense of their weave and pattern. From this we understand why artworks are superior to drugs in cleansing perception. Though they cannot produce amphetamine's euphoria, they make up for it at an intellectual level by putting objects and events in context. They hint at, if not explicitly state, the relatedness—the larger pattern, perception of which is not only mentally but also sensorially rewarding. As I noted earlier, to Wordsworth a daisy is not just a flower in isolated glory but part of a larger system that includes the sun, a blade of grass, and the dewdrop. Indeed, the daisy owes its singularity and moral grandeur to its position among these other objects; relative size, location, and distance all have consequence.

Artworks, argues Iris Murdoch, can make one a better person, if only because they teach quiet, self-forgetful attentiveness.[54] Unfortunately, too often this virtue of attentiveness is locked into the world of art and is not readily transferable to things and humans beyond; in other words, people tend to treat the poem, picture, or music, with all the internal relatedness and complexities that give it fascination, as a sealed rather than as a perforated reality.

Without doubt, focusing on the pattern diminishes the richness of its particulars. As one admires the harmony of the heavenly spheres, the individual stars fade. As one tries to understand the system of cities, the cities themselves become abstract points and circles, exercising varying degrees of socioeconomic pull and push. As one strives to envisage a communal ideal—the obligations and duties of one person to another—the individual member can no longer be an end in itself, except perhaps at a transient moment, when the entire communal "dance" pivots on it. This diminishment of the particular in all its concrete specificity arouses unease, for we are such particulars. To those who take the utmost delight in the sheer plenitude and variety of dearly beloved objects in creation, including individual human beings, physical science is especially suspect, for it pushes abstraction to the point where relatedness is all and the particulars are almost nothing.

Ashes and Dust

Yet the nothingness of the particular, including the human individual, is a key doctrine of the world's great religions and philosophies. The British artist who intended to prick the bloated self-esteem of modern men and women by saying, over and over again, "We are just meat!" was not more harsh than the Bible, which teaches that we are just creatures of the day like "grass" (Psalm 90), or the Anglican Book of Common Prayer, which reminds us at every funeral service that our bodies are but "ashes and dust." Buddhism is perhaps most extreme in this regard. A Buddhist monk is exhorted to look at the human body as "an old bag with orifices at top and bottom, and stuffed full of groceries, encased with skin and full of impure matter."[55] As for the soul, it has no permanent existence; indeed, the Buddha taught that belief in such an entity is among the most powerful of human delusions.[56]

Modern science at its best is the great austere religion of our time. It teaches necessity. It teaches the insignificance of the particular. Both are hard doctrines that every scientist accepts. Science consoles *because* it offers bracing necessity rather than delusory hope. Einstein was one among many who shunned philosophies that pander to exaggerated ideas of human freedom. It was evident to him that "everybody acts not only under external compulsion but also in accordance with inner necessity." Schopenhauer's saying, "A man can do what he wants, but not want what he wants," was an inspiration to Einstein from youth onward; it provided him with an unfailing wellspring of comfort and patience amid the hardships of life, his own and others'.[57] Einstein never tired of reminding himself of his own insignificance in the larger scale of things, and, by implication, the insignificance of other human individuals as well. During an awesome storm on December 10, 1930, he felt as though he—this tiny particle of existence—had simply "dissolved and merged into Nature." And he was happy.[58] There is something Buddhist, or just deeply religious, in his response.

B. F. Skinner provides another example of the religious temperament. Skinner religious? It may seem a strange idea, for he is widely known as a hard-nosed behavioral scientist. Yes, but I would argue that being hard-nosed is precisely that which should earn him the religious epithet. Skinner himself would not disagree. Let me explain. Most people like to see themselves as free—only a little lower than the angels, creators as God is creator. Skinner, by contrast, sees people as essentially the playthings of genetic factors and environmental history. That is the scientific position. But Skinner reminds us that this scientific position is also "an act of self-denial that would have been understood by Thomas à Kempis." He quotes Luke (17.33): "Whoever seeks to gain his life will lose it, but whoever loses his life will preserve it," and goes on to say that this idea, congenial to the ethos of science, is embraced "in mundane philosophy (Schopenhauer's annihilation of the will as the way to freedom) and literature (Conrad's Secret Sharer learning that true self-possession comes from self-abandonment)." And "it is, of course, a strong theme in Eastern mysticism." Skinner was interviewed upon the publication of his autobiography. He concluded the interview with a remark that is as religious-ascetic as it is scientific: "If I am right about human behavior, I have written the autobiography of a nonperson."[59]

"I am a nonperson" is a healthy point of view only if it is the result of self-discovery. To be thus labeled by others is, obviously, not healthy but annihilating. Discovering and concluding that I am of little import presupposes privilege, the opportunity and ability to appreciate the vast canvas of space and time on which impersonal forces rule; for it is against such background awareness that I see my own small part as not only vanishingly small but largely unfree. Note that Skinner does not deny freedom altogether; he just wants the tininess of its scope recognized. He may even admit that at a psychological level such recognition can open the door to adventure and true discovery. Religious thinkers have always urged that an individual consider his own life disposable, for it is only when he is willing to lay it aside that he can hope to regain it at a higher rung. By contrast, the insistent boast that "I am somebody" can delimit and limit the claimant, locking him or her into that somebody—that heavy mix of flesh and status—he or she happens to be. Existentialist philosophy has something to say here too. It postulates not a core essence but a void at the human center. Unless that void is filled too soon or injudiciously—a recurrent temptation—it permits, even compels, the human individual to take up the world. In time, the world—the universe—is everything and the self happily almost nothing. In religious language, one has been absorbed, though not annihilated, by the divine. One is in heaven.

God vs. Neighbors: Final Escape to the Real

Confucianism is primarily a humanist doctrine because its central concern is with the quality of human relationships in this world. Nevertheless, it is infused by two other kinds (and conceptions) of reality that are far older than Confucianism itself. One is a sense of the holy. The other is the belief that community extends beyond the living to the dead—to ancestors—and to the unborn. Among world religions, Christianity is exceptional in its emphasis on human relationships, the heights to which they can rise as distinct from the number of rules and regulations governing them. Christianity is almost Confucian in this regard, as is its doctrine of the communion of saints, which extends membership beyond the living to the dead. Jesuit missionaries in the sixteenth and seventeenth centuries successfully proselytized the Chinese by directing attention to such commonalities.[60] Christianity, however, is not by any stretch a humanism. One has only to look at its prescriptions for conduct, which are so extreme that they must be deemed unrealistic if intended for earthbound creatures. They make sense only in a religious context, in which one's neighbor is made in the image of God, in which the sick man, the poor woman, or the prisoner in jail is Christ. Christianity is also neither a humanism nor a natural religion because it teaches that our final home is not on earth but elsewhere. It directs attention heavenward. It is out of step with the natural and chthonic religions of the past, as it is with the various nostalgic earth cults of the present. As understood by early Christians—an understanding that has not totally disappeared with the passage of time—their first duty is not to other people either, nor to the endless maintenance tasks that make human society possible. The greatest commandment is not "Love your neighbor as yourself" but "Love the Lord your God with all your heart, with all your soul, and with all your mind" (Matthew 22.37–39). In the story of Martha and Mary, it was Martha who invited Jesus to her home and did the hospitable thing by working in the kitchen; it was Mary her sister who sat at the Lord's feet and listened to his teaching. When Martha complained, the Lord answered her, "Martha! Martha! You are worried and troubled over so many things, but just one is needed. Mary has chosen the right thing, and it will not be taken away from her" (Luke 10.38–42).

What is the real and good? What makes for happiness? One answer is the one that Creon gives to his niece Antigone in Jean Anouilh's play by that name: "Life is a child playing around your feet, a tool you hold firmly in your grip, a bench you sit on in the evening, in the garden. . . . Believe me, the only poor consolation that we have in our old age is to discover that what I have said to you is true." Sounds reasonable, doesn't it? What else is there other than these solidly real, endearing particulars? Pass them by and chase the wind, and one will end up with—well, the wind. Yet Antigone's response to her uncle is one of outrage: "I spit on your happiness! I spit on your idea of life."[61]

Missing from Creon's picture are two elements that he would surely have wanted to add if his thoughts had moved in that direction: charity to neighbors and respect for nature. Thus completed, this picture of the good life—so human and humane, practical and realizable—has exerted consistent appeal to people past and present, including modern man and woman in a sober mood. It is what one expects in our time from wise tribal elders, sensitive secular humanists, and concerned environmentalists. Yet not only the little spitfire Antigone but also thinkers both Eastern and Western, steeped in their highly developed religious/philosophical traditions, would have nothing to do with it. Or rather, they would see it as secondary to other longings.

Earlier I noted as strange the fact that humans exist who make the world they live in far more frightful than it is by populating it with a host of malicious but nonexistent beings—this strangeness lingering even when we can see that such feverish turns perform a service by objectifying deep personal anxieties and fears, locating them outside the self and so making them seem more manageable. But from the viewpoint of common sense, isn't it even stranger that, historically and today, some of the world's best and most serious minds have concluded that the concrete particulars—the sensory delights and joys of everyday life, all the things that obviously matter, including sustainable agriculture and the care of the earth—are provisional, are shadows and illusions, or, at best, hints and intimations; and that the really real, the ultimate goal of all attempts at escape, is the One, the Void (or Nirvana), the Good, the Beatific Vision, the Sun that blinds, or even Steven Weinberg's final unifying theory of nature, which is expected to be of incomparable beauty?[62]

Wanting to reach that goal can seem selfish. The ascetic hermit-monk of early Buddhism, bent on accumulating merits so as to escape another embodied existence in this world, is not a sympathetic figure. Nor is the utterly dedicated scientist who neglects not only his own personal needs but also those of his family and friends, absent-mindedly telling them to wait as he reaches for the greater beauty behind yet another veil. What about charity, helping those in material need, and human fellowship? Aren't these the greater values, the greater deeds? Eastern religions and philosophies, in essence, say no. Their ultimate sights are set elsewhere and do not include community service. Indeed, Taoism is explicitly amoral. Since transcendent Tao considers all distinctions of right and wrong, true and false, as irrelevant, what ground is there for virtuous, self-sacrificing conduct?[63] In this regard, Buddhism appears to be the outstanding exception among Eastern philosophies. But is it? The famous compassion for all living things in Buddhism derives not from the ability to put oneself in the position of another, or from an all-encompassing nature philosophy of interdependence, but from a system of metaphysics at the center of which is transmigration. It is not only that the chicken one eats for dinner may be one's grandma, but also that all living things strive and so needlessly suffer; in compassion, then, one seeks to relieve the suffering born of desire by becoming oneself an example of desirelessness. The Bodhisattvas who choose not to enter Nirvana do not become social workers; rather, they sit, their eyes closed and turned inward, images of perfect peace, models of nonstriving—the only way out of the endless operations of karma.[64] In the West, the ancient Greeks and Romans had no conception of any moral duty to help those less fortunate. Christianity did and does emphasize such service, but, as I have just noted, even there the contemplative life has always ranked higher than the active one.

In the great religious and philosophical traditions, compassionate voices can be heard condemning individuals who ignore the neighbors they can see for a God they cannot see. Such a path is condemned out of genuine sympathy for those in need who are thus sidetracked, but also because it is all too easy to follow and involves, moreover, the contradiction of seeking spiritual values selfishly. So the well-off and competent are called upon to feed and care for the poor and the feeble, the wounded and the sick, and enjoined to remember that they themselves are, in one sense or another, to varying degree, poor, feeble, wounded, and sick. Nevertheless, from the viewpoint of the great traditions, these duties are more like chores of housekeeping than ultimate ends. On earth, everything has to be maintained and periodically repaired, whether this be a house, the human body, or the social body with its egregious imbalances. It comes as a jolt to realize that this cannot be true of heaven. By definition heaven has no poor and sick, no exploited class, and so no housekeeping, no material caring, no revolutionary fervor to right wrong, none of those things that make up the bulk of ethical life on earth and that, furthermore, give individual human beings their sense of belonging, virtue, and importance. In heaven, helping others remains central and is more devotedly practiced than ever, but it can only be of a spiritual-intellectual kind, as senior angels show their juniors the further delights of knowing God.

What is heaven like? Given our earthbound nature, we are likely to envisage it in terms of particulars. Particulars may be excellent in themselves, but we with our feeble and needy imagination are too liable to turn them into idols. So it may be the better part of wisdom to envisage God and heaven in terms of great abstractions and negativities, which, while they may have a powerful aesthetic/emotional appeal—the restfulness of the Void, the beauty of mathematical equations—discourage idol-worshiping and the incurable human yearning to possess.

What, in the end, is the great escape? One of my favorite stories is that concerning Thales of Miletus. He fell into a well while gazing upward doing astronomy. A clever Thracian serving-girl is said to have made fun of him—a man so eager to know the things in heaven that he failed to notice what was right next to his feet.[65] Thales tasted both the heights and the depths, while the Thracian girl moved along the safe, horizontal plane. Who had the better part? Can anyone have the better part, the ultimate escape of Thales and Mary? I think so, for it is not a matter of talent, or even of socioeconomic circumstance, but of a willingness to look in the right direction.

Notes

• • •

1 Earth / *Nature and Culture*

1. An earlier version of this chapter appeared as "Escapism: Another Key to Cultural-Historical Geography," *Historical Geography* 25 (1997): 10–24. I thank editors Steven Hoelscher and Karen Till for permission to use it.

2. Christopher Stinger and Clive Gamble, *In Search of the Neanderthals: Solving the Puzzle of Human Origins* (New York: Thames and Hudson, 1993).

3. Albert Hirschman, *Exit, Voice, and Loyalty* (Cambridge: Harvard University Press, 1970).

4. "Every beginning is difficult," says Goethe. An Australian historian applies this dictum to his own country: "In Australia, every beginning has not only been difficult, but scarred with human agony and squalor." In C. M. H. Clark, *Select Documents in Australian History, 1851–1900* (Sydney: Augus and Robertson, 1955), 94. For a grim account of frontier life in the United States, see Everett Dick, *The Lure of the Land: A Social History of the Public Lands from the Articles of Confederation to the New Deal* (Lincoln: University of Nebraska Press, 1970).

5. Colin Turnbull, *Wayward Servants* (London: Eyre and Spottiswode, 1965), 20–21.

6. Inga Clendinnen, *Aztec: An Interpretation* (Cambridge: Cambridge University Press, 1995), 29–32.

7. For the self-confidence and optimism of the Chinese, even during the Shang dynasty, see David N. Keightley, "Late Shang Divination," in *Explorations in Early Chinese Cosmology*, ed. Henry Rosemont Jr. (Chico, Calif.: Scholars Press, 1984), 22–23.

8. Benjamin Schwartz, *The World of Thought in Ancient China* (Cambridge: Harvard University Press, 1985); Herbert Fingarette, *Confucius: The Secular As Sacred* (New York: Harper Torchbooks, 1972). For an example of an imperial memorial to heaven, see S. Wells Williams, *The Middle Kingdom*, rev. ed. (New York: Charles Scribner's Sons, 1907), 1:467–68.

9. Pierre Goubert, *Louis XIV and Twenty Million Frenchmen* (New York: Pantheon Books, 1970), 178–81, 216.

10. For a vivid example of the uncertainties of life in seventeenth-century England, see the story of the clergyman-farmer and his family in Alan Macfarlane, *The Family Life of Ralph Josselin, a Seventeenth-Century Clergyman* (Cambridge: Cambridge University Press, 1970).

11. Stephen Orgel, *The Illusion of Power: Political Theater in the English Renaissance* (Berkeley: University of California Press, 1975), 51–55.

12. Alfred North Whitehead, *Science and the Modern World* (New York: Mentor Books, 1959), 42–43.

13. N. K. Sandars, *The Epic of Gilgamesh* (Harmondsworth, Middlesex: Penguin, 1964), 30–31.

14. Mary Douglas, "The Lele of Kasai," in *African Worlds*, ed.

Daryll Forde (London: Oxford University Press, 1963), 1–26.

15. Robert C. Ostergren, *A Community Transplanted: The Trans-Atlantic Experience of a Swedish Immigrant Settlement in the Upper Middle West* (Madison: University of Wisconsin Press, 1988).

16. William Cronon, ed., *Uncommon Ground: Toward Reinventing Nature* (New York: Norton, 1995); Neil Evernden, *The Social Creation of Nature* (Baltimore: Johns Hopkins University Press, 1992).

17. Paul Engelmann, ed., *Letters from Ludwig Wittgenstein, with a Memoir* (Oxford: Blackwell, 1967), 97–99.

18. Gillian Gillison, "Images of Nature in Gimi Thought," in *Nature, Culture, and Gender*, ed. Carol MacCormack and Marilyn Strathern (Cambridge: Cambridge University Press, 1980), 144.

19. For the concept of boundary among the Mbuti Pygmies of the Congo (Zaire) forest, see Colin Turnbull, *The Mbuti Pygmies: An Ethnographic Survey*, Anthropological Papers of the American Museum of Natural History, vol. 50, pt. 3 (New York: American Museum of Natural History, 1965), 165.

20. Marilyn Strathern, "No Nature, No Culture: The Hagen Case," in *Nature, Culture, and Gender*, ed. MacCormack and Strathern, 174–222; see also J. R. Goody, *The Domestication of the Savage Mind* (Cambridge: Cambridge University Press, 1977).

21. Victor Turner, *The Ritual Process: Structure and Anti-Structure* (Ithaca: Cornell University Press, 1969).

22. Mircea Eliade, *The Sacred and the Profane: The Nature of Religion* (New York: Harper Torchbooks, 1961).

23. "Middle landscape" is an eighteenth-century idea that became a powerful tool for understanding the people-environment relationship in the second half of the twentieth century, thanks to Leo Marx. See his *The Machine in the Garden: Technology and the Pastoral Ideal in America* (New York: Oxford University Press, 1964), 100–103.

24. Yi-Fu Tuan, "Gardens of Power and Caprice," in *Dominance and Affection: The Making of Pets* (New Haven: Yale University Press, 1984), 18–36.

25. John M. Findlay, "Disneyland: The Happiest Place on Earth," in *Magic Lands: Western Cityscapes and American Culture after 1940* (Berkeley: University of California Press, 1992), 56–116.

2 Animality / *Its Covers and Transcendence*

1. Larissa MacFarquhar, "The Face Age: Can Cosmetic Surgery Make People into Works of Art?" *New Yorker*, July 21, 1997, 68–70.

2. Note the ambiguous meaning of the word "human" itself. We tend to forget that "human" is of the same root as "humus" and "humility."

3. Edward Moffat Weyer, *The Eskimos: Their Environment and Folkways* (New Haven: Yale University Press, 1932), 72.

4. Colin Turnbull, "The Lesson of the Pygmies," *Scientific American*, January 1963, 1–11; Kevin Duffy, *Children of the Forest* (New York: Dodd, Mead and Co., 1984), 161–66.

5. Colin Thubron, *Behind the Wall: A Journey through China* (London: Heinemann, 1987), 182–84.

6. For a sparkling account of a Roman eating orgy, see Petronius, "Dinner with Trimalchio," in *The Satyricon*, trans. William Arrowsmith (New York: Mentor Books, 1960), 38–84.

7. Simon Schama, "Mad Cows and Englishmen," *New Yorker*, April 8, 1996, 61.

8. Nick Fiddes, *Meat: A Natural Symbol* (London and New York: Routledge, 1991), 16.

9. Ethnographic examples: Mary Douglas, "The Lele of Kasai," in *African Worlds*, ed. Daryll Forde (London: Oxford University Press, 1963), 1–26; Gillian Gillison, "Images of Nature in Gimi Thought," in *Nature, Culture, and Gender*, ed. Carol MacCormack and Marilyn Strathern (Cambridge: Cambridge University Press, 1980), 143–73; M. Shostak, *Nisa: The Life and Words of a !Kung Woman* (Harmondsworth, Middlesex: Penguin, 1983); N. Chagnon, *Yanomamo: The Fierce People* (London: Holt, Rinehart and Winston, 1977), 29, 33.

10. Stephen Mennell, *All Manners of Food: Eating and Taste in England and France from the Middle Ages to the Present* (Oxford: Blackwell, 1987), 31–32.

11. Charles Cooper, *The English Table in History and Literature* (London: Sampson Low, Marston and Co., n.d.), 3.

12. K. C. Chang, ed., *Food in Chinese Culture: Anthropological and Historical Perspectives* (New Haven: Yale University Press, 1977), 7–10.

13. Ibid., 37–38

14. E. N. Anderson, *The Food of China* (New Haven: Yale University Press, 1988), 114.

15. Confucius, "Lun Yu," in *The Four Books,* trans. James Legge (New York: Paragon Reprints, 1966), 130.

16. Frederick W. Mote, "Yuan and Ming," in *Food in Chinese Culture,* ed. Chang, 238.

17. Jasper Griffin, *Homer on Life and Death* (Oxford: Clarendon Press, 1986), 19–20.

18. Richard Sennett, *The Fall of Public Man* (Cambridge: Cambridge University Press, 1975), 182.

19. Joseph R. Levenson, *Liang Ch'i-Ch'ao and the Mind of Modern China* (Berkeley: University of California Press, 1970), 117–18.

20. Philip Gourevitch, "Letter from Rwanda: After the Genocide," *New Yorker,* December 18, 1995, 78–94.

21. Jacques J. Maquet, *The Promise of Inequality in Ruanda: A Study of Political Relations in a Central African Kingdom* (London: Oxford University Press, 1961), 10, 18–19. I draw on his work in the next two paragraphs.

22. Poem cited in Joseph Levenson and Franz Schurmann, *China: An Interpretive History* (Berkeley: University of California Press, 1971), 114–15.

23. Daniel A. Dombrowski, *The Philosophy of Vegetarianism* (Amherst: University of Massachusetts Press, 1984), 19–74.

24. Caroline Walker Bynum, *Holy Feast and Holy Fast: The Religious Significance of Food to Medieval Women* (Berkeley: University of California Press, 1987), 33–47.

25. True, a scholar-artist's studio in East Asia could seem immaculate, but that only signified the greater need for—and greater cunning of—cover.

26. Allan Bloom, *Love and Friendship* (New York: Simon and Schuster, 1993), 45.

27. Jane Goodall, *The Chimpanzees of Gombe: Patterns of Behavior* (Cambridge: Harvard University Press, 1986), 138.

28. Ibid., 447–48.

29. C. S. Lewis, "Prudery and Philology," in *Present Concerns*, ed. Walter Hooper (San Diego: Harcourt Brace Jovanovich, 1986), 88–89.

30. Quoted in Lionel Trilling, *Sincerity and Authenticity* (Cambridge: Harvard University Press, 1972), 5.

31. B. Karlgren, "Some Fecundity Symbols in Ancient China," *Bulletin, Museum of Far Eastern Antiquities* (Stockholm), no. 2 (1930): 1–21.

32. Otto J. Brendell, "The Scope and Temperament of Erotic Art in the Greco-Roman World," in *Studies in Erotic Art*, ed. Theodore Bowie and Cornelia V. Christenson (New York: Basic Books, 1970), 12.

33. David Brion Davis, "Slaves in Islam," *New York Review of Books*, October 11, 1990, 36. The book reviewed is Bernard Lewis, *Race and Slavery in the Middle East: An Historical Enquiry* (New York: Oxford University Press, 1990).

34. Ashley Montagu, *Touching: The Human Significance of the Skin* (New York: Harper and Row, 1978).

35. Rollo May, *Love and Will* (New York: W. W. Norton, 1969), 75.

36. Doris Lessing, *The Golden Notebook* (New York: Simon and Schuster, 1962), 479–80.

37. John Liggett, *The Human Face* (New York: Stein and Day, 1974).

38. Roger Scruton, *Sexual Desire: A Moral Philosophy of the Erotic* (New York: Free Press, 1986), 26, 150–51, 154.

39. Irving Singer, *The Pursuit of Love* (Baltimore: Johns Hopkins University Press, 1994), 19–20.

40. Francesco Alberoni, *Falling in Love* (New York: Random House, 1983), 29, 30, 35.

41. William Jankowiak, ed., *Romantic Passion: A Universal Experience?* (New York: Columbia University Press, 1995).

42. Niklas Luhmann, *Love As Passion: The Codification of Intimacy* (Cambridge: Harvard University Press, 1986), 21.

43. Octavio Paz, *The Double Flame: Love and Eroticism* (New York: Harcourt Brace and Co., 1995), 19.

44. Ananda K. Coomaraswamy, *The Arts and Crafts of India and Ceylon* (New York: Noonday Press, 1964), 65.

45. Kenneth Clark, *The Nude: A Study in Ideal Form* (Princeton: Princeton University Press, 1990), 307.

46. Robert Bernard Martin, *Gerard Manley Hopkins: A Very Private Life* (New York: Putnam's, 1991), 114.

47. Peter Brown, *The Body and Society: Men, Women, and Sexual Renunciation in Early Christianity* (New York: Columbia University Press, 1988), 47–48, 53–64.

48. Marina Warner, *Alone of All Her Sex: The Myth and the Cult of the Virgin Mary* (New York: Vintage Books, 1983), 54–55.

49. Virginia Woolf, *The Diary of Virginia Woolf*, ed. Anne Olivier Bell (New York: Harcourt Brace Jovanovich, 1980), 3:117.

50. Epicurus, "Letter to Menoeceus." See Cyril Bailey, trans., *Epicurus: The Extant Remains* (Oxford: Oxford University Press, 1926).

51. Teilhard de Chardin, *The Divine Milieu: An Essay on the Interior Life* (New York: Harper, 1960).

52. Quoted in Burton Watson, *Chinese Lyricism: Shih Poetry from the Second to the Twelfth Century* (New York: Columbia University Press, 1971), 49–50.

53. Iris Murdoch, *The Sovereignty of Good* (New York: Schocken Books, 1971), 99.

54. Alexander Alland Jr., *Adaptation in Cultural Evolution: An Approach to Medical Anthropology* (New York: Columbia University Press, 1970), 160.

55. Jorge Luis Borges, *Twenty-four Conversations with Borges, Including a Selection of Poems: Interviews, 1981–1983, by Roberto Alifano* (Housatonic, Mass.: Lascaux Publishers, 1984), 4.

56. *New York Times*, February 24, 1984.

57. Sidney Hook, "Pragmatism and the Tragic Sense of Life," in *Contemporary American Philosophy*, ed. John E. Smith (London: Allen and Unwin, 1970), 179.

58. Karl R. Popper and John C. Eccles, *The Self and Its Brain* (Heidelberg, London, and New York: Springer International, 1981), 556.

59. John Cowper Powys, *The Art of Happiness* (London: John Lane, Bodley Head, 1935), 46–47, 74.

60. Oliver St. John Gogarty, "To Death," in *The Collected Poems of Oliver St. John Gogarty* (New York: Devin-Adair, 1954), 191.

61. Malcolm Muggeridge, *Jesus Rediscovered* (New York: Doubleday, 1969), 106.

62. Karl Barth, *The Task of the Ministry*, quoted in John Updike, *Assorted Prose* (New York: Knopf, 1965), 282.

63. James Boswell, *Life of Samuel Johnson* (Chicago: Encyclopaedia Britannica, 1952), 394.

64. Quoted in W. Jackson Bate, *Samuel Johnson* (New York: Harcourt Brace Jovanovich, 1979), 451–52.

65. S. N. Kramer, *The Sumerians* (Chicago: University of Chicago Press, 1963), 263.

66. *Odyssey* 24.5ff. See C. M. Bowra, *The Greek Experience* (New York: Mentor Books, 1957), 50–52.

67. Dan Davin, "Five Windows Darken: Recollections of Joyce Cary," *Encounter*, June 1975, 33.

68. E. O. James, *The Beginnings of Religion: An Introductory and Scientific Study* (London: Hutchinson's University Library, 1950), 129.

69. Knud Rasmussen, *Intellectual Culture of the Iglulik Eskimos*, Report of the Fifth Thule Expedition, 1921–24, vol. 7, no. 1 (Copenhagen: Gyldendalske Boghandel, Nordisk Forlag, 1929), 73–75.

70. John W. Berry, "Temne and Eskimo Perceptual Skills," *International Journal of Psychology* 1 (1966): 207–29.

71. As an example of the dreary life and architecture in the underworld, see Emily M. Ahern, *The Cult of the Dead in a Chinese Village* (Stanford: Stanford University Press, 1973).

72. See Dorothy Sayers's characterization of Dante's paradise in *Essays Presented to Charles Williams*, ed. C. S. Lewis (Grand Rapids, Mich.: Eerdmans, 1966), 30–31; E. J. Becker, *A Contribution to the Comparative Study of the Medieval Visions of Heaven and Hell* (Baltimore: John Murphy, 1899); and Colleen McDannell and Bernhard Lang, *Heaven: A History* (New Haven: Yale University Press, 1988).

73. Irene Masing-Delic, *Abolishing Death: A Salvation Myth of Russian Twentieth-Century Literature* (Stanford: Stanford University Press, 1992).

3 People / *Disconnectedness and Indifference*

1. Stuart F. Spicker, ed., *The Philosophy of the Body* (Chicago: Quadrangle Books, 1970); Martin Buber, *I and Thou* (New York: Scribner's, 1958); Paul Vanderbilt, *Between the Landscape and Its Other* (Baltimore: Johns Hopkins University Press, 1993).

2. Dorothy Lee, "Linguistic Reflection of Wintu Thought" and "The Conception of the Self among the Wintu Indians," in *Freedom and Culture* (Englewood Cliffs, N.J.: Prentice-Hall, 1959), 121–30, 130–40; Y. P. Mei, "The Individual in Chinese Social Thought," in *The Status of the Individual in East and West*, ed. Charles A. Moore (Honolulu: University of Hawaii Press, 1968), 333–48; Bruno Snell, *The Discovery of the Mind: The Greek Origins of European Thought* (Cambridge: Harvard University Press, 1953), 60; Yi-Fu Tuan, *Segmented Worlds and Self: Group Life and Individual Consciousness* (Minneapolis: University of Minnesota Press, 1982), 82, 139–67.

3. This chapter draws on my previous paper, "Island Selves: Human Disconnectedness in a World of Indifference," *Geographical Review* 85, no. 2 (1995): 229–39. I thank the American Geographical Society for permission to use it.

4. Roger Williams, *You Are Extraordinary* (New York: Random House, 1967); "Nutritional Individuality," *Human Nature*, June 1978, 46–53.

5. M. Neitz and J. Neitz, "Numbers and Ratios of Visual Pigment Genes for Normal Red-Green Color Vision," *Science* 267 (February 17, 1995): 1013–18.

6. Jacques Hadamard, *The Psychology of Invention in the Mathematical Field* (Princeton: Princeton University Press, 1949), 115.

7. M. S. Gazzaniga, *Nature's Mind: The Biological Roots of Thinking, Emotions, Sexuality, Language, and Intelligence* (New York: Basic Books, 1992).

8. John Updike, *Self-Consciousness* (New York: Knopf, 1989), 105.

9. Elizabeth Bowen, "Ivy Gripped the Steps," in *Collected Stories* (London: Jonathan Cape, 1980), 707–8.

10. Jules Henry, *Pathways to Madness* (New York: Random House, 1971), 88.

11. Albert Camus, *Carnets, 1942–1951* (London: Hamish Hamilton, 1966), 37.

12. For an original exposition of the idea of the world's indifference, see L. Kolakowski, *The Presence of Myth* (Chicago: University of Chicago Press, 1989).

13. Harold Nicolson, *The War Years, 1939–1945* (New York: Atheneum, 1967), 30.

14. Clarence J. Glacken, *Traces on the Rhodian Shore: Nature and Culture in Western Thought from Ancient Times to the End of the Eighteenth Century* (Berkeley: University of California Press, 1967), 375–428.

15. One example out of many is Vicki Hearne's account of the sensitivity and intelligence of horses. See her *Adam's Task: Calling Animals by Name* (New York: Knopf, 1986).

16. Roderick Nash, *The Rights of Nature: A History of Environmental Ethics* (Madison: University of Wisconsin Press, 1989); Neil Evernden, *The Social Creation of Nature* (Baltimore: Johns Hopkins University Press, 1992).

17. Karl A. Nowotny, *Beiträge zur Geschichte des Weltbildes: Farben and Weltrichtungen*, Wiener Beiträge zur Kulturgeschichte und Linguistik,

vol. 17 (1969), (Vienna: Verlag Ferdinand Berger & Sohne, 1970). For two case studies from opposite ends of the earth, see Alfonso Ortiz, *New Perspectives on the Pueblos* (Albuquerque: University of New Mexico Press, 1972), and John B. Henderson, *The Development and Decline of Chinese Cosmology* (New York: Columbia University Press, 1984).

18. C. S. Lewis, *A Preface to Paradise Lost* (London: Oxford University Press, 1960), 22–31. I can't resist offering another story of indifference—a particularly haunting one—from our own time: Four friends went on a yachting trip in the Mediterranean. It was a warm, calm day. The sea was as smooth as glass. The yacht drifted to a stop. Before taking lunch, the friends decided to jump into the sea for a swim. They splashed about happily and, having worked up an appetite and no doubt thinking of the chicken and wine on the cabin table, they decided to climb aboard. To their horror, they found they couldn't. The boat rose too high above water and since it was unanchored, the swimmers had no rope to haul themselves up with. They all drowned. I see in my mind's eye the swimmers in their last desperate minutes, wondering, as their arms turned to lead, at the utter indifference of serene sky and sea. This real-life tragedy was told to Robert Stone, the novelist, who described it as the "the ultimate 'oops.'" I embellished the story for my own purpose. See Bruce Weber, "An Eye for Danger," *New York Times Magazine*, January 19, 1992, 19.

19. I have in mind desired or desirable bodily contact as part of daily living. In modern times, bodily contact can actually be crushing and stressful, as, for example, during the rush hour in a jam-packed car of a New York subway train or in Japanese mass transit. A reviewer of the manuscript drew my attention to this point.

20. Jules Henry, *Jungle People: A Kaingang Tribe of the Highlands of Brazil* (New York: J. J. Augustin, 1941), 18, 33; Colin Turnbull, "The Ritualization of Potential Conflict between the Sexes among the Mbuti," in *Politics and History in Band Societies*, ed. E. Leacock and R. Lee (Cambridge: Cambridge University Press, 1982), 137.

21. Victor Zuckerkandl, *Man the Musician* (Princeton: Princeton University Press, 1973).

22. Colin Turnbull, "Liminality: A Synthesis of Subjective and Objective Experience," in *By Means of Performance: Intercultural Studies of Theatre and Ritual*, ed. Richard Schechner and Willa Appel (Cambridge: Cambridge University Press, 1990), 56.

23. Zuckerkandl, *Man the Musician*, 27–28.

24. J. Glenn Gray, *The Warriors: Reflections on Men in Battle* (New York: Harper Torchbooks, 1967), 45.

25. James William Gibson, *Warrior Dreams: Violence and Manhood in Post-Vietnam America* (New York: Hill and Wang, 1994), 108–9.

26. Denis Wood and Robert J. Beck, *Home Rules* (Baltimore: Johns Hopkins University Press, 1994).

27. The most rigorous and extended analysis of the relationship of place to people is in Robert David Sack, *Homo Geographicus: A Framework for Action, Awareness, and Moral Concern* (Baltimore: Johns Hopkins University Press, 1997), 60–126.

28. Yi-Fu Tuan, "Language and the Making of Place," *Annals of the Association of American Geographers* 81, no. 4 (1991): 684–96, and "The City and Human Speech," *Geographical Review* 84, no. 2 (1994): 144–51.

29. B. B. Whiting and J. W. M. Whiting, *Children of Six Cultures: A Psycho-Cultural Analysis* (Cambridge: Harvard University Press, 1975), 170–71.

30. George Steiner, "The Language Animal," *Encounter*, August 1969, 7–24.

31. James Fernandez, "The Mission of Metaphor in Expressive Culture," *Current Anthropology* 15 (1974): 119–45.

32. Hans Jonas, *The Phenomenon of Life: Toward a Philosophical Biology* (New York: Harper and Row, 1966), 11–12.

33. John Bayley, ed., *The Portable Tolstoy* (New York: Viking, 1978), 37–38. Wittgenstein, who read Tolstoy diligently, might have been influenced by the Russian, as the following passage suggests: "What makes a subject hard to understand—if it's something significant and

important—is not that before you can understand it you need to be specially trained in abstruse matters, but the contrast between understanding the subject and what most people *want* to see. Because of this the very things which are most obvious may become the hardest of all to understand. What has to be overcome is a difficulty having to do with the will, rather than with the intellect." In *Culture and Value* (Chicago: University of Chicago Press, 1980), 17.

34. Reported by Julian Green in *Diary, 1928–1957* (New York: Carroll and Graf, 1985), 63. Because of large biological differences, a man may not be able to communicate with his dog at a deep level. No such large differences exist among people. Yet despite the possession of common speech, they too often fail to communicate, and the reason for failure this time would seem to be egotism—a moral defect in human beings.

35. Ilham Dilman, *Love and Human Separateness* (Oxford: Blackwell, 1987).

36. In the major universities, from the 1960s onward, the intellectual disposition of graduate students in the social sciences and humanities has tended to be left of center and Marxist.

37. Quoted in Barre Toelken, *The Dynamics of Folklore* (Boston: Houghton Mifflin, 1979), 96. On the intimate relationship between landscape and self, both group and individual, see Leslie Marmon Silko, "Interior and Exterior Landscapes: The Pueblo Migration Stories," in *Landscape in America*, ed. George F. Thompson (Austin: University of Texas Press, 1995), 155–69.

38. Jean L. Briggs, *Aspects of Inuit Value Socialization*, National Museum of Man Mercury Series, Canadian Ethnology Service Paper no. 56 (Ottawa: National Museum of Canada, 1979), 6.

39. John Updike, *Rabbit Is Rich* (New York: Knopf, 1981), 116.

40. Knud Rasmussen, *Intellectual Culture of the Iglulik Eskimos*, Report of the Fifth Thule Expedition, 1921–24, vol. 7, nos. 2 and 3 (Copenhagen: Gyldendalske Boghandel, Nordisk Forlag, 1930), 19, 69.

41. Ibid., 59.

42. J. Drury, *Angels and Dirt* (New York: Macmillan, 1974), 52.

43. W. H. Auden, "Death's Echo," in *Collected Poems*, ed. E. Mendelson (New York: Vintage Books, 1991), 153.

44. Iris Murdoch, *A Word Child* (London: Chatto and Windus, 1975), 45.

45. Claude Lévi-Strauss and D. Erbon, *Conversations with Claude Lévi-Strauss* (Chicago: University of Chicago Press, 1991), 102–3.

46. Denis Cosgrove and Stephen Daniels, "Iconography and Landscape," in *The Iconography of Landscape*, ed. Denis Cosgrove and Stephen Daniels (Cambridge: Cambridge University Press, 1988), 1–10.

47. Aldous Huxley, "Unpainted Landscapes," *Encounter*, October 1962, 41–47.

48. Arthur C. Danto, *The Philosophical Disenfranchisement of Art* (New York: Columbia University Press, 1986), 89–91; E. H. Gombrich, *Art and Illusion: A Study in the Psychology of Pictorial Representation* (London: Phaidon Press, 1962), 9–12.

49. W. H. Auden, "Musée des Beaux Arts," in *A Selection by the Author* (Harmondsworth, Middlesex: Penguin, 1958), 61.

4 Hell / *Imagination's Distortions and Limitations*

1. Hans Jonas, *Mortality and Morality: A Search for the Good after Auschwitz*, ed. Lawrence Vogel (Evanston: Northwestern University Press, 1996), 13.

2. Norman Cohen, *Europe's Inner Demons: An Enquiry Inspired by the Great Witch-Hunt* (New York: Basic Books, 1975), 70.

3. George Santayana, *Reason in Society*, vol. 2 of *The Life of Reason* (1905; reprint, New York: Dover Publications, 1980), 81.

4. Colin Wilson, *Origins of the Sexual Impulse* (London: Arthur Baker, 1963), 167.

5. Wilhelm von Humboldt, *Humanist without Portfolio* (Detroit: Wayne State University Press, 1963), 383–84.

6. J. Glenn Gray, *The Warriors: Reflections on Man in Battle* (New York: Harper Torchbooks, 1967), 51.

7. For a provocative interpretation of cruelty, see Clement Rosset, *Joyful Cruelty: Towards a Philosophy of the Real* (New York: Oxford University Press, 1993).

8. Elizabeth Marshall Thomas, *The Harmless People* (New York: Vintage Books, 1965), 51–52.

9. *Terre Inhumaine*, quoted in Albert Camus, *Carnets, 1942–1951* (London: Hamish Hamilton, 1966), 142.

10. John Updike, *Self-Consciousness* (New York: Knopf, 1989), 150. The rationalization of torture as a device for obtaining truth is, of course, also beyond animals. See Edward Peters, *Torture* (Philadelphia: University of Pennsylvania Press, 1996).

11. Wolfgang Sofsky, *The Order of Terror: The Concentration Camp* (Princeton: Princeton University Press, 1996), 225–26, 237.

12. This example is taken from anthropologist-photographer Kevin Duffy's book, *Children of the Forest* (New York: Dodd, Mead and Co., 1984), 50. He is a more recent observer of the Mbuti Pygmy scene than Colin Turnbull, a widely recognized authority.

13. The distinguished writer V. S. Naipaul was born in Trinidad. "It's really wonderful that we no longer laugh at people with disabilities in Trinidad," he writes. "Black people once laughed at people's disabilities. It was very cruel. I remember the black audience at the Port-of-Prince cinema when the concentration camps were uncovered in Germany at the end of the war. The black audience, you know, *shocking me* by laughing at the inmates in the newsreels." In *Conversations with V. S. Naipaul*, ed. Feroza Jussawalla (Jackson: University Press of Mississippi, 1997), 120.

14. Colin Turnbull, *The Forest People* (Garden City, N.Y.: Doubleday, 1962), 100.

15. Nelson Graburn, "Severe Child Abuse among the Canadian Inuit," in *Child Survival: Anthropological Perspectives on the Treatment and Mal-*

treatment of Children, ed. N. Scheper-Hughes (Dordrecht: D. Reidel, 1987), 211–26; Michael Baksh, "Cultural Ecology and Change of the Machiguenga Indians of the Peruvian Amazon" (Ph.D. diss., University of California, Los Angeles, 1984), 99–100. I owe both sources to Robert B. Edgerton, *Sick Societies: Challenging the Myth of Primitive Harmony* (New York: Free Press, 1992), 79–80.

16. I owe this insight to Frederic Cassidy, a linguist at the University of Wisconsin–Madison. See Yi-Fu Tuan, "Community and Place: A Skeptical View," in *Person, Place, and Thing*, ed. S. T. Wong, Geoscience and Man, vol. 31 (Baton Rouge: Geoscience Publications, Dept. of Geography and Anthropology, Louisiana State University, 1992), 50. Limited aggression actually contributes to community. See David Gilmore, *Aggression and Community* (New Haven: Yale University Press, 1987).

17. Quoted in James A. Knight, *For the Love of Money: Human Behavior and Money* (Philadelphia: Lippincott, 1968), 161.

18. J. C. Speakman, *Molecules* (New York: McGraw-Hill, 1966), vi.

19. James William Gibson, *Warrior Dreams: Violence and Manhood in Post-Vietnam America* (New York: Hill and Wang, 1994), 109–12.

20. Jean-Paul Sartre, *Saint Genet* (New York: Braziller, 1963), 360–61.

21. David McCullough, *Mornings on Horseback* (New York: Simon and Schuster, 1981), 88.

22. Enid Welsford, *The Fool: His Social and Literary History* (London: Faber and Faber, n.d.), 135; E. Tietze-Conrat, *Dwarfs and Jesters in Art* (London: Phaidon Press, 1957), 80.

23. Cao Xueqin, *The Story of the Stone* (Harmondsworth, Middlesex: Penguin, 1980), vol. 3, *The Warning Voice*, 157.

24. Frank E. Huggett, *Life below Stairs: Domestic Servants in England from Victorian Times* (London: John Murray, 1977), 27.

25. Barrington Moore Jr., *Injustice: The Social Bases of Obedience and Revolt* (White Plains, N.Y.: M. E. Sharpe, 1978), 55–64.

26. See cartoon and accompanying text on p. 200 of John W. Dower, *War without Mercy: Race and Power in the Pacific War* (New York: Pantheon, 1986).

27. Lucretius, *On the Nature of Things*, bk. 2, trans. Ronald Latham (Harmondsworth, Middlesex: Penguin, 1951), 60. For a general account of the aesthetics of destruction—from predators serenely squashing their prey out of existence (Thoreau) to the beauty of shipwreck—see Philip Hallie, *Tales of Good and Evil, Help and Harm* (New York: HarperCollins, 1997), 118–29.

28. Lynn White Jr., "Death and the Devil," in *The Darker Vision of the Renaissance*, ed. Robert S. Kinsman (Berkeley: University of California Press, 1974), 32.

29. *Time Magazine*, December 13, 1976, 78.

30. The morality (and immorality) of witnessing a public execution was vigorously debated by Russian writers. See Robert Louis Jackson, *Dialogues with Dostoevsky: The Overwhelming Questions* (Stanford: Stanford University Press, 1993).

31. Tim Cresswell, *In Place/Out of Place: Geography, Ideology, and Transgression* (Minneapolis: University of Minnesota Press, 1996); Robert David Sack, *Human Territoriality: Its Theory and History* (Cambridge: Cambridge University Press, 1986), and *Homo Geographicus: A Framework for Action, Awareness, and Moral Concern* (Baltimore: Johns Hopkins University Press, 1997), 90–91, 156–60.

32. Tepilit Ole Saitoti, *The Worlds of a Maasai Warrior* (Berkeley: University of California Press, 1988), 73.

33. Sack, *Homo Geographicus*, 32, 127–41.

34. Fung Yu-lan, *A Short History of Chinese Philosophy* (New York: Macmillan, 1959), 71–72. Mencius: "Now, does E really think that a man's affection for the child of his brother is merely like his affection for the infant of a neighbor?" This is Mencius's response to E's observation that "we are to love all without difference of degree." In Mencius, bk. 3, chap. 5. As Benjamin Schwartz puts it, "Mencius believes that the

very love of mankind in general is only possible as a result of the irradiation—in diminishing strength, to be sure—of that love which naturally has its source in the bosom of the family." In *The World of Thought in Ancient China* (Cambridge: Harvard University Press, 1985), 259.

35. Speaking of his experience with various tribes in Gabon while working with Albert Schweitzer in Lambaréné, Frederick Frank writes, "Because of the easy rapport I establish with people of the various tribes, I found their intertribal relationships all the more puzzling. For them, people from another tribe have little claim to sympathy. The good Samaritan is only conceivable if you are a Samaritan yourself. There is no concept of universal brotherhood. A man has fallen out of a palm tree. If he is from your tribe you carry him on your back to the next village. If he is not, you pass by and let him die." In *My Days with Albert Schweitzer* (New York: Lyons and Burford, n.d.), 97; see also 81–82. As for American Indians, the Mashantucket Pequot tribe of Connecticut, which operates a prosperous casino, is one of the richest in the nation, but it makes no effort to help its poorer brethren. Instead it counts on white foundations to dispense charity and largesse. See Paul Boyer, "Poor Little Rich Indians," *Tribal College: Journal of American Indian Higher Education*, Winter 1994–95, 4–5.

36. Lawrence Stone, review of Barbara Tuchman, *A Distant Mirror: The Calamitous Fourteenth Century*, in *New York Review of Books*, September 28, 1978, 4.

37. Walt Whitman, "Song of Myself," st. 51, in *Leaves of Grass* (New York: Signet Classics, 1960), 96.

38. Here is another grim, less well known example of compartmentalization: "The Einsatzkommando operating in the region of Simferopol, inside Russia, was ordered to kill three thousand Jews and Gypsies before Christmas. The order was carried out with exceptional speed so that the troops could attend the celebration of Christ's birth." In Tzvetan Todorov, *Facing the Extreme: Moral Life in the Concentration Camps* (New York: Henry Holt, 1996), 148–49.

39. Robert Hutchins, "The Scientist's Morality," *Minority of One*, November 1963, 25.

40. Primo Levi, *The Drowned and the Saved* (New York: Summit Books, 1988), 57–58.

41. E. M. Forster, *Howards End* (New York: Vintage Books, 1989), 195.

42. Humphrey Osmond, "Schizophrenics and Empathy," *Mental Hospitals*, Architectural Supplement, 8 (April 1957): 23–30.

43. J. Robert Oppenheimer, "Prospects in the Arts and Sciences," in *Man's Right to Knowledge* (New York: Columbia University Press, 1954), 114–15.

44. Helen Vendler, *The Odes of John Keats* (Cambridge: Harvard University Press, 1983), 26, 85.

45. Anthony Heilbut, *Thomas Mann: Eros and Literature* (New York: Knopf, 1996), 21–22.

46. Whitman, "Song of Myself," st. 32, in *Leaves of Grass*, 73.

47. Roger Shattuck, *The Forbidden Experiment: The Story of the Wild Boy of Aveyron* (New York: Farrar Straus Giroux, 1980), 182.

48. Michael Ignatieff, "His Art Was All He Mastered," *New York Times Book Review*, August 29, 1988, 24.

49. George Orwell, *Down and Out in Paris and London* (New York: Berkley, 1959), 16.

50. Ferdynand Zweig, *The Quest for Fellowship* (London: Heinemann, 1965), 132.

51. "To a Skylark" by Percy Bysshe Shelley is the 140th most anthologized poem in the English language. See *The Top Five Hundred Poems*, ed. William Harmon (New York: Columbia University Press, 1992), 1081. The poem itself appears on p. 500.

52. Shattuck, *Forbidden Experiment*, 180; Marjorie Hope Nicolson, *Voyages to the Moon* (New York: Macmillan, 1948).

53. Robert Wohl, *A Passion for Wings: Aviation and the Western Imagination, 1908–1918* (New Haven: Yale University Press, 1994).

54. Antoine de Saint-Exupéry, *Wind, Sand, and Stars* (Harmondsworth, Middlesex: Penguin, 1966), 24.

55. Reported in *Minneapolis Star and Tribune*, June 13, 1983, and in *Time Magazine*, November 4, 1996, 80.

56. Umberto Eco, *Travels in Hyperreality* (New York: Harcourt Brace Jovanovich, 1990); Jean Baudrillard, *America* (New York: Verso Press, 1988). See also Greil Marcus, "Forty Years of Overstatement: Criticism and the Disney Theme Parks," in *Designing Disney Theme Parks: The Architecture of Reassurance*, ed. Karal Ann Marling (Montreal: Canadian Centre for Architecture, 1997), 201–7.

57. Robert Darnton, "The Meaning of Mother Goose," *New York Review of Books*, February 2, 1984, a review of *The Types of Folk-Tale: A Classification and Bibliography*.

58. Yi-Fu Tuan with Steven D. Hoelscher, "Disneyland: Its Place in World Culture," in *Designing Disney Theme Parks*, ed. Marling, 191–98.

5 Heaven / *The Real and the Good*

1. Consider this much quoted passage from Einstein: "The scientist's religious feeling takes the form of rapturous amazement at the harmony of natural law, which reveals an intelligence of such superiority that compared with it, all systematic thinking and acting of human beings is an utterly insignificant reflection." In *Ideas and Opinions* (New York: Random House, 1948), 43.

2. Marjorie Hope Nicolson, *The Breaking of the Circle* (New York: Columbia University Press, 1962).

3. John G. Neihardt, *Black Elk Speaks* (Lincoln: University of Nebraska Press, 1961), 198.

4. Stephen Toulmin, *Foresight and Understanding: An Enquiry into the Aims of Science* (New York: Harper Torchbooks, 1963).

5. Lynn White Jr., *Machine ex Deo* (Cambridge: MIT Press, 1968), 17.

6. Andrew Hodges, *Alan Turing: The Enigma* (New York: Simon and Schuster, 1983), 207.

7. G. J. Sussman and Jack Wisdom, "Chaotic Evolution of the Solar System," *Science* 257 (July 3, 1992): 56–62; Ray Jayawardhana, "Tracing

the Milky Way's Rough-and-Tumble Youth," *Science* 258 (November 27, 1992): 1439.

8. S. Chandrasekhar, *Truth and Beauty: Aesthetics and Motivations in Science* (Chicago: University of Chicago Press, 1987), 54.

9. Roger Penrose, *The Large, the Small, and the Human* (Cambridge: Cambridge University Press, 1997).

10. Stephen Hall, "News and Comments," *Science* 259 (March 12, 1993): 1533.

11. Robert E. Murphy, *The Dialectics of Social Life: Alarms and Excursions in Anthropological Theory* (New York: Basic Books, 1971), 126–27.

12. Philippe Ariès, *Centuries of Childhood: A Social History of Family Life* (New York: Vintage Books, 1965).

13. James Fernandez, "The Mission of Metaphor in Expressive Culture," *Current Anthropology* 15, no. 2 (1974): 122–23.

14. Paul Chance, "The Me I Didn't Know," *Psychology Today*, January 1987, 20.

15. For *feng-shui* in Hong Kong, see Veronica Huang, "Hong Kong's Tower of Assorted Trouble," *Wall Street Journal*, October 12, 1976, 1, and Stephen Skinner, *The Living Earth Manual of Feng-Shui* (London: Arkana/Penguin, 1989), 32–33, 130–31. For *feng-shui* in Singapore and Malaysia, see Derham Groves, *Feng-Shui and Western Building Ceremonies* (Singapore: Tynron Press, 1991), 14–17, 19–23.

16. There is a vast literature on progress, whether it is real. See Raymond Duncan Gastil, *Progress: Critical Thinking about Historical Change* (Westport, Conn.: Praeger, 1993).

17. Bertrand Russell is reported to have said that physics is mathematical not because we know so much about the physical world but because we know so little; it is only its mathematical properties that we can discern. Reported by Arthur Koestler in *The Act of Creation* (New York: Macmillan, 1964), 251.

18. Philip P. Hallie, *The Paradox of Cruelty* (Middletown, Conn.: Wesleyan University Press, 1969), 46.

19. Richard Schechner and Willa Appel, eds., *By Means of Performance: Intercultural Studies of Theatre and Ritual* (Cambridge: Cambridge University Press, 1990). Here is Susanne Langer's way of saying that dance enchants: "In dance, powers become apparent in a framework of space and time. . . . Dance's true creative aim—to make the world of Powers visible." Why does dance matter when it is no longer wrapped in religion and has lost its magical purposes? Langer's answer is this: "The eternal popularity of dance lies in its ecstatic function, today as in earliest times; but instead of transporting the dancers from a profane to a sacred state, it now transports them from what they acknowledge as 'reality' to a realm of romance." In *Feeling and Form: A Theory of Art* (New York: Charles Scribner's Sons, 1953), 199, 201–2.

20. Howard Gardner, *Art, Mind, and Brain* (New York: Basic Books, 1982), 86, 93, 97–100.

21. William Wordsworth, "To a Child Written in Her Album," in *The Poems*, ed. John O. Hayden (London: Penguin, 1989), 2:780.

22. Newton P. Stallknecht, *Strange Seas of Thought: Studies in William Wordsworth's Philosophy of Man and Nature* (Bloomington: Indiana University Press, 1958).

23. Gardner, *Art, Mind, and Brain*, 88–90.

24. S. Honkaavara, *The Psychology of Expression*, British Journal of Psychology Monograph Supplements, no. 32 (Cambridge: Cambridge University Press, 1961).

25. Aldous Huxley, "Unpainted Landscapes," *Encounter* 19 (1962): 41–47.

26. Jay Appleton, *The Experience of Landscape*, rev. ed. (Chichester: John Wiley, 1996).

27. Robert D. Sack, *Conceptions of Space in Social Thought* (London: Macmillan, 1980); Yi-Fu Tuan, *Space and Place: The Perspective of Experience* (Minneapolis: University of Minnesota Press, 1977).

28. There is a large and excellent literature on landscape. See, for example, Kenneth R. Olwig, "Recovering the Substantive Nature of

Landscape," *Annals of the Association of American Geographers* 86, no. 4 (1996): 630–53; George F. Thompson, ed., *Landscape in America* (Austin: University of Texas Press, 1995); Denis Cosgrove, *The Palladian Landscape* (University Park: Pennsylvania State University Press, 1993); Denis Cosgrove and Stephen Daniels, eds., *The Iconography of Landscape* (Cambridge: Cambridge University Press, 1988); Simon Schama, *Landscape and Memory* (New York: Knopf, 1995); and J. B. Jackson, *Discovering the Vernacular Landscape* (New Haven: Yale University Press, 1984).

29. The preeminence of music in Chinese and European civilization is well known. In China, music was believed to emanate from heaven and to provide harmony not only in human society but in the entire cosmos. A Confucian scholar-official must not only know *li*, propriety and the rules of conduct, but also *yueh*, music. See Li Ki, "Yueh Chi," bk. 17, in *The Sacred Books of the East* (Oxford: Clarendon Press, 1884), 28:115, and Bliss Wiant, *The Music of China* (Hong Kong: Chung Chi Publications, Chinese University of Hong Kong, n.d.), 7. As for Europe, the relation of cosmic harmony to music—the idea that the harmony of the spheres is also the music of the spheres—goes back to the Pythagoreans and remained influential in learned circles until the seventeenth century. George Steiner reminds us of the ancient belief that music had the power to build cities: "The magus . . . can make stones of music. One version of the myth has it that the walls of Thebes was built by song." See Steiner, "The City under Attack," in *The Salmagundi Reader*, ed. Robert Boyers and Peggy Boyers (Bloomington: Indiana University Press, 1983), 4. Among great Christian thinkers, Augustine believed that poetry, even the poetry of the Psalms, marked a fall into imperfection and temporality. But the fall was less catastrophic in music than in other arts "because music is comparatively free of the superstition that reality is entirely what it appears to be, finite, bodily, visible, and therefore imitable." Music, in other words, remains heavenly, out of this world, invisible. See Denis Donoghue, *Warrenpoint* (New York: Knopf, 1990), 82. To our contemporary, the literary critic and polymath George Steiner, "the creative spirit reaches its utmost perfection in music: to him, music is time 'made organic'; it is,

quite simply, 'the supreme mystery of man.'" Gerhard Neuman, "George Steiner's Real Presences," in *Reading George Steiner*, ed. Nathan A. Scott Jr. and Ronald A. Sharp (Baltimore: Johns Hopkins University Press, 1994), 253. Elsewhere in the same book Steiner draws attention to the great anthropologist "[Claude] Lévi-Strauss's arresting formulation: the invention of melody . . . remains the *mystère suprême des sciences de l'homme*" (284).

30. Jamie James, *The Music of the Spheres: Music, Science, and the Natural Order of the Universe* (New York: Copernicus, 1993).

31. Think of how great artists have repeatedly tried to capture an aspect of the real world: Rembrandt with his ceaseless effort to capture the self in self-portraits, Monet with his nine almost identical Mornings on the Seine, Shih-t'ao with his seventy-two views of a single mountain, the Huang Shan. There is nothing remotely like this in music.

32. Jacques Barzun, "Is Music Unspeakable?" *American Scholar*, Spring 1996, 196.

33. Gerald Brenan, *Thoughts in a Dry Season* (Cambridge: Cambridge University Press, 1978), 77.

34. Peter Kivy, *Music Alone: Philosophical Reflections on the Purely Musical Experience* (Ithaca: Cornell University Press, 1990).

35. R. Murray Schafer, *The Tuning of the World* (New York: Knopf, 1977), 115–18, 156.

36. David Plante, "Bacon's Instinct," *New Yorker*, November 1, 1993, 96.

37. A neat example is provided by the five cultures of northwestern New Mexico: "The self-image of each cultural group is ethnocentrically flattering. All refer to themselves in terms that define them as 'persons' or 'people,' and to the others as excluded from or inferior to true humanity. The Navahos call themselves *dineh*, which literally means 'the people,' and the Zuni call themselves *ashiwi*, literally 'the flesh' or the 'the cooked ones.' The Mormons, in keeping with their Old Testament restoration, have adopted the ancient Hebraic title, 'the

chosen people.' More tolerant in their view of outsiders, the Spanish-Americans nevertheless think of themselves as *la gente*, 'people,' in the honorific sense. The Texans share the self-congratulatory vision of the 'real' or 'super' American, figuratively the only really 'white men.'" In Evon Vogt and Ethel Albert, eds., *People of Rimrock: A Study of Values in Five Cultures* (Cambridge: Harvard University Press, 1966), 26.

38. On community and place at different scales, see J. Nicholas Entrikin, *The Betweenness of Place: Towards a Geography of Modernity* (Baltimore: Johns Hopkins University Press, 1991), 62–83.

39. Karl Jaspers, *The Origin and Goal of History* (New Haven: Yale University Press, 1953), 1–21.

40. *Analects* xvii.2, quoted in Donald J. Munro, *The Concept of Man in Early China* (Stanford: Stanford University Press, 1969), 13.

41. Arthur Waley, *The Analects of Confucius* (London: Allen and Unwin, 1983), bk. 16, chap. 9, p. 206.

42. Benjamin Schwartz, *The World of Thought in Ancient China* (Cambridge: Harvard University Press, 1985), 83.

43. Herbert Fingarette, *Confucius: The Secular As Sacred* (New York: Harper Torchbooks, 1972), 1–5. Fingarette uses the expression "holy rite" rather than "holy dance," and he interprets "holy" as magic, a word that I studiously avoid.

44. Aristotle was the reasonable man. By contrast, Socrates was in his way as scandalous as Jesus. In fact, Socrates and Jesus had much in common in their refusal of retribution—their rejection of the law of the talion. See George Steiner, *No Passion Spent* (Chicago: University of Chicago Press, 1996), 380–81.

45. The story of the prodigal son is bound to raise Confucian eyebrows.

46. Leslie Fiedler, "The Rebirth of God and the Death of Man," in *The Salmagundi Reader*, ed. Robert Boyers and Peggy Boyers (Bloomington: Indiana University Press, 1983), 376.

47. J. H. Hexter, *The Judaeo-Christian Tradition*, 2d ed. (New Haven:

Yale University Press, 1995); Page Smith, *Rediscovering Christianity: A History of Modern Democracy and the Christian Ethic* (New York: St. Martin's Press, 1994).

48. Peter J. Wilson, *The Domestication of the Human Species* (New Haven: Yale University Press, 1988).

49. Quoted in Victor Gollancz, ed., *A Year of Grace* (Harmondsworth, Middlesex: Penguin, 1955), 75–76.

50. J. R. R. Tolkien, "On Fairy-Stories," in *Essays Presented to Charles Williams*, ed. C. S. Lewis (Grand Rapids, Mich.: Eerdmans, 1966), 44–45.

51. C. N. Manlove, "On the Nature of Fantasy," in *The Aesthetics of Fantasy Literature and Art*, ed. Roger C. Schlobin (Notre Dame: University of Notre Dame Press, 1982), 30–31. Modern fairy tales are moralistic; not so the fantastic folk tales of the seventeenth and eighteenth centuries upon which they may be based. See Robert Darnton, "The Meaning of Mother Goose," *New York Review of Books*, February 2, 1984, 45–46.

52. C. S. Lewis's science fiction is a notable exception. In the volumes of his trilogy, *Out of the Silent Planet* (1938), *Perelandra* (1943), and *That Hideous Strength* (1945)—aptly characterized as his works of theological science fiction—good and evil are at the heart of the story.

53. Aldous Huxley, *The Doors of Perception and Heaven and Hell* (New York: Perennial Library, 1990), 20.

54. Iris Murdoch, *The Sovereignty of Good* (New York: Schocken Books, 1971), 64–65, 69, 85–88.

55. G. F. Allen, *The Buddha's Philosophy: Selections from the Pali Canon and an Introductory Essay* (New York: Macmillan, 1959), 162–63.

56. "The Buddha was known as the Anatta-vadi, or teacher of Impersonality." In *Buddhist Dictionary—Manual of Buddhist Terms and Doctrines: Nyanatiloka*, Island Hermitage Publications 1 (Colombo, Sri Lanka: Frewin and Co., 1950); Edward Conze, *Buddhist Thought in India* (Ann Arbor: University of Michigan Press, 1967), 122–34.

57. Albert Einstein, *The World As I See It* (New York: Covici, Friede, 1934), 238.

58. Helen Dukas and Banesh Hoffmann, *Albert Einstein: The Human Side* (Princeton: Princeton University Press, 1989), 23.

59. Interview with B. F. Skinner in *Psychology Today*, September 1983, 30, 32.

60. *China in the Sixteenth Century: The Journals of Matthew Ricci, 1583–1610*, trans. Louis J. Gallagher (New York: Random House, 1953).

61. Jean Anouilh, *Antigone and Eurydice: Two Plays* (London: Methuen, 1951), 56, 58. For a critical evaluation of immortal longings beyond the humdrum good that Creon recommends to Antigone, see Fergus Kerr, *Immortal Longings: Versions of Transcending Humanity* (Notre Dame: University of Notre Dame Press, 1997).

62. Steven Weinberg, *Dreams of a Final Theory: The Scientist's Search for the Ultimate Laws of Nature* (New York: Vintage Books, 1994).

63. Herlee G. Creel, *What Is Taoism?* (Chicago: University of Chicago Press, 1970), 3.

64. Arthur C. Danto, *Mysticism and Morality: Oriental Thought and Moral Philosophy* (New York: Columbia University Press, 1987), 26, 39, 77.

65. Plato, *Theaetetus* 174a.

Acknowledgments

• • •

Since this may be my last academic work, I would like to thank all who helped me in a long career. All? Of course, that's not possible. I can't even remember the name of my nurse who taught me the basic Chinese characters. So, not all. Nevertheless, I welcome the opportunity to acknowledge my indebtedness to as many as space here allows.

First, my mentors of half a century ago: R. P. Beckinsale (Oxford), Clarence Glacken, John Kesseli, John Leighly, Jim Parsons, Erhard Rostlund, and Carl Sauer (Berkeley). Then, my long list of friends and colleagues who have gone out of their way to be supportive, among whom I include Arlin Fentem (Indiana), B. L. Gordon, J. B. Jackson (New Mexico), Joe May (Toronto), Michael Steiner (CSU), John Borchert, John Fraser Hart, Helga Leitner, Fred Lukermann, Cotton Mather, Roger Miller, Philip Porter, Eric Sheppard (Minnesota), Paul Claval (Sorbonne), Nick Entrikin (UCLA), David Lowenthal (UCL), and Peter Gould (Penn State). And everyone at Wisconsin! I do mean everyone, yet I can't resist naming the Big Five: Jim Knox, Bob Ostergren, Sharon Ruch, Bob Sack, and Tom Vale. Beyond Geography, at Wisconsin, I wish to thank Leon Berkowitz, Lloyd Bitzer, David and Jean Cronon, Merle Curti, Betsy Draine, Don Emmerson, Emiko Ohnuki-Tierney, Francis Schrag, and Don Smith for their sympathetic interest. "Sympathetic interest" may not sound like a powerful endorsement, but it means much to me. Then, students who have taken the trouble to tell me "where it is at": Mark Bouman, Philippe Cohen, Michael Curry, Linda Graber, John Hickey, Patrick McGreevy, Kenneth Olwig, April Veness (Minnesota); Paul Adams, Tim Bawden, Tom Boogaart II, Tim Cresswell, Steven Hoelscher, Carol Jennings, Matthew Kurtz, Eric Olmanson, Jemuel Ripley, Karen Till, Richard Waugh, Jeff Zimmerman, and members of the seminars on "Escapism" (Wisconsin).

Behind every paper or book is the editor. I've had good editors, dedicated editors, but the one who has given me the most consistent support and encouragement in the last fifteen years is the former editor of the *Geographical Review*, Douglas R. McManis. As for *Escapism*, it owes its appearance in print to the unflagging support of George Thompson, president of the Center for American Places and an editor for the Johns Hopkins University Press. In my curmudgeonly old age, I fight hard against obscurity in writing—almost a fashion among younger scholars—in favor of lucidity; and in this battle I couldn't have a better ally than my copy editor, Mary Yates.

Last but not least, I wish to express my gratitude to two rooms: 548 Social Sciences Building, University of Minnesota, and 243 Science Hall, University of Wisconsin. I have come to regard these rooms as my brain. Outside them, I am a hollow man. Inside them, I can think and write. The spirit of place deserves much credit (and blame) for my papers and books; and if this sounds like a displacement of responsibility, I suggest that it is allowable in a geographer.

Illustration Credits

• • •

pp. 2–3 "View of Mount Holyoke, Northampton, Massachusetts, after a Thunderstorm (The Oxbow)," by Thomas Cole, Metropolitan Museum of Art, New York, gift of Mrs. Russell Sage.

p. 29 "Princess X," by Constantin Brancusi, Philadelphia Museum of Art: The Louise and Walter Arensberg Collection.

pp. 78–79 "Many Happy Returns of the Day," by William Powell Frith, Mercer Art Gallery, Harrogate.

p. 111 "Saturn Devouring His Son," by Francisco de Goya y Lucientes, Museo del Prado, Madrid. Image from Giraudon/ Art Resource, New York.

p. 151 "Jacob's Ladder," by William Blake, The British Museum, London.

Index

• • •

Library of Congress Cataloging-in-Publication Data
Tuan, Yi-fu, 1930–
Escapism / Yi-Fu Tuan.
p. cm.
Includes bibliographical references and index.
ISBN 0-8018-5926-3 (hardcover : alk. paper)
1. Escape (Psychology) 2. Nature. 3. Culture. I. Title.
BF575.E83T83 1998
153.3—dc21 98-5366
CIP

Milton Keynes UK
Ingram Content Group UK Ltd.
UKHW031846251024
450150UK00001B/15